North Sea Bonanza

by Peter Fairley

Hart-Davis

Mobil

Part 1
Riches Under the Waves

On July 29th 1959 a drill belonging to the NAM Company of Holland bit through hard rock 2600 metres underground near the Dutch town of Gröningen and released a jet of gas. It was colourless and odourless – mostly methane or 'marsh' gas – a natural gas formed by the decay of forests and other vegetation which had become buried with time.

Scientists had suspected it. Engineers had searched for it, for years. But it lay deeper than anticipated. When they finally found it, they soon realized that they had located a natural treasure of enormous value.

Other holes were drilled. The Gröningen discovery turned out to be one of the largest gasfields in the world and will meet much of Holland's energy need well into the 21st Century.

As the gas began to flow, geologists reasoned that the same rock formation might well extend west, beyond Holland and out into the North Sea. Similar rock formations had already been found in Yorkshire. They carried out underwater surveys. They quickly confirmed that the whole of the North Sea was really two great 'basins' whose bottom layers might not only harbour gas but oil as well.

Meanwhile, the governments of nations bordering the North Sea – Britain, France, Belgium, Holland, Germany, Denmark, Sweden and Norway – had been busy.

The whole of the North Sea is less than 200 metres deep and counts – under international law – as Continental Shelf. The seabed belongs, therefore, to the country or countries adjoining it. By the mid-1960s, a series of agreements had been drawn up between the eight governments which apportioned the underwater territory fairly between them.

Any oil company wishing to prospect for gas or oil had to apply for a licence from the government concerned. Any oil or gas found could not be extracted without a *further* licence and payment of tax.

Drilling began. In October 1965 came the first 'strike' – natural gas lying in what is now called the West Sole Field. Further 'strikes' followed. It was soon apparent that the southern basin of the North Sea was an enormous reservoir for gas which had seeped out of ancient coal seams.

Prospectors moved further north. They found that the second of the North Sea basins – divided from the first by a mound of rock known as the Northumbrian Arch and extending between Scotland and Norway – was rich in oil.

The first oil 'strike' was made in 1970. Since then, 17 major oil fields have been discovered. The first oil was landed in Norway in 1974 and in Britain in 1975.

Today, an armada of some 200 ships, drilling rigs and helicopters move restlessly about the North Sea, engaged in the business of bringing the oil and gas ashore, or hunting for more.

Giant towers of steel and islands of concrete have sprung up, marking the places where production is in full flow. Thousands of men are pitting their wits and their brawn against one of the wildest stretches of ocean in the world, to exploit Nature's gift.

For us ashore, the North Sea bonanza is just beginning. The economic miracle is yet to come. But already the people of Britain, of Holland, of Norway and of Denmark – those so far lucky enough to have made a North Sea 'strike' – are beginning to feel a change in their lives . . .

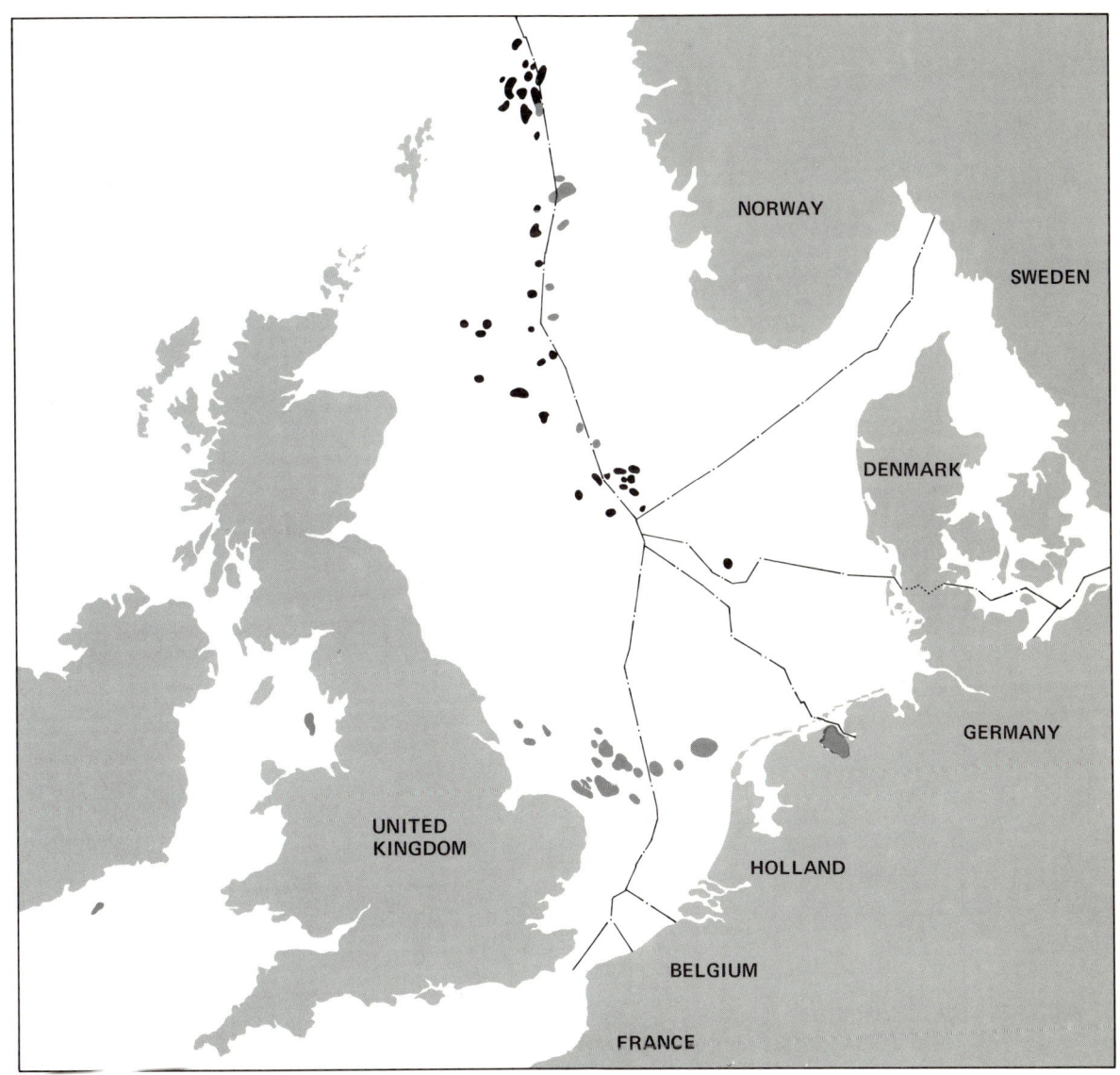

North Sea oilfields (black) and gasfields (shaded) discovered so far. Lines indicate sectors 'owned' by different nations.

Part 2
How the Oil and Gas Were Formed

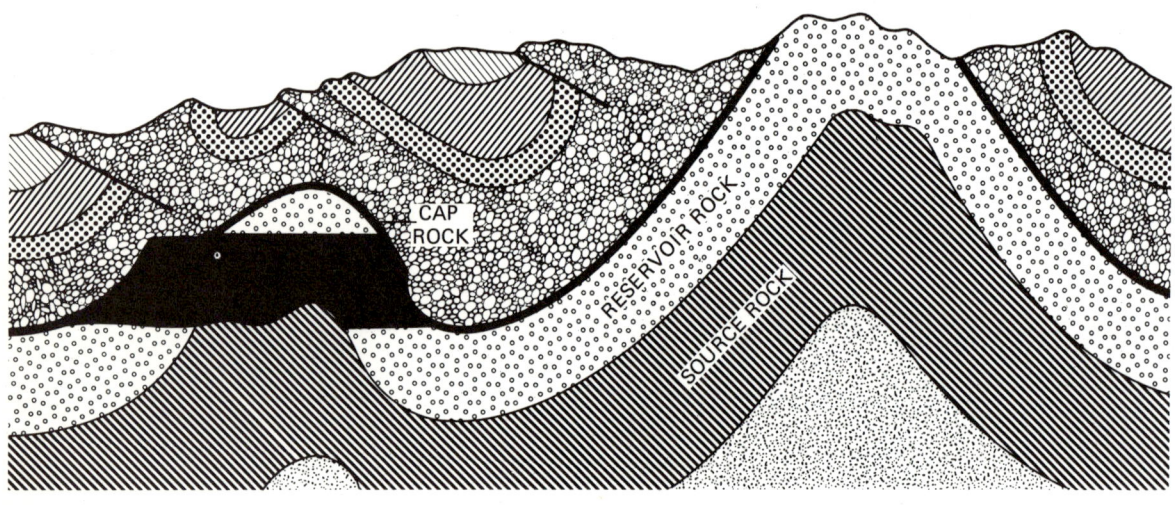

How the Earth's crust crinkles, allowing oil to pool (thin black line represents hard cap rock).

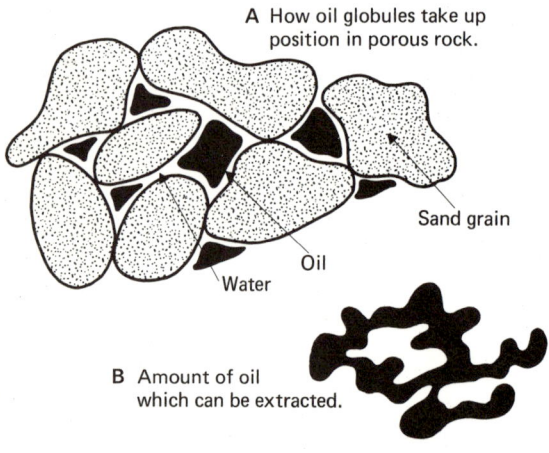

A How oil globules take up position in porous rock.

Sand grain

Oil

Water

B Amount of oil which can be extracted.

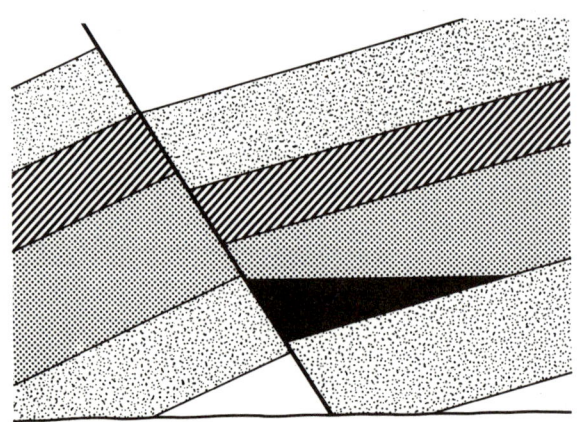

▲ *Typical rock fault allowing oil (black) to pool.*

North Sea oil was once alive – or so most scientists believe.

They think it was formed out of millions of tiny organisms – marine creatures and plants – which died, sank to the bottom and decomposed. Down in the murky depths, hundreds of millions of years ago, the decaying matter was gradually buried by particles eroded from nearby land and carried down to the sea by rivers.

Bacteria went to work. Minerals in the mud acted as catalysts. The fatty acids in the organisms were changed into hydrocarbons. Gradually droplets of oil were formed, not always black, but slippery . . .

As more sediment built up, layer upon layer, so the pressure and temperature increased. The oil was squeezed. It moved about.

It sought space in the pores of harder rock. It migrated upwards, being lighter than the water driving it about. Eventually it reached impermeable rock – 'cap' rock – and came to rest.

At the same time, the Earth's crust was being subjected to tremendous forces. In places it heaved and contorted. The rocks crinkled. Some folded, others slipped into faults. The sliding and slipping left underground traps into which the oil could pool. And there it has lain ever since, truly a 'fossil fuel', waiting for man to find it.

Gas in the North Sea is thought to have been formed rather differently. Although a quantity of gas is often found in the same area as oil, the so-called 'natural' gasfields in the North Sea are believed to have been formed by seepage out of deeply buried coal seams.

Four geological conditions are necessary for an area to contain oil or gas:

Source Rock: This is the rock on which the droplets of oil were originally formed out of organic matter. The most common source rocks are black shales or bituminous limestones.

Reservoir Rock: This is a porous rock – such as sandstone – through which the oil molecules can move freely. The more permeable the reservoir rock, the fewer wells needed for its draining.

Cap Rock: This seals the oil below the sea-bed. The most common cap rocks are salts and shales which can build up into thicknesses of several thousand metres.

Folds or faults: These enclose the oil or gas-field on all sides to make it worthwhile extracting the fuel. Generally speaking, not enough oil or gas is formed by natural decay to make it worth drilling except in places where the rocks have crinkled and allowed the oil to pool.

Once the drill has penetrated, crude oil comes to the surface as a mixture of hundreds of different chemicals, ranging from extremely light gases to semi-solid, burnable materials such as asphalt or paraffin. The 'mix' varies from place to place in the North Sea.

The natural gas found so far in the North Sea is remarkably 'clean' – free from sulphur and other pollutants – and so requires a minimum of processing and treatment before being piped to users. That makes it economically attractive.

Part 3
Prospecting

Oil and gas have lain under the surface of the Earth for millions of years. But it was not until 1859 that men began to drill for it.

The first oil well was sunk by 'Colonel' Edwin Drake at Titusville, Pennsylvania. For many years, oil exploration remained a haphazard business – drilling was done mainly in places where oil had obviously seeped through to the surface, or in 'wildcat' fashion, at random. But gradually geologists built up an understanding of rock formations which favour oil and gas and the search became more scientific.

Four different techniques of prospecting are now used in the North Sea.

First geologists prepare maps and cross-sectional drawings of what they *believe* the rocks below to be like. Possible source rocks are listed. Porous rocks are indicated and the thickness of cap rocks estimated.

The geologists then hand over to the geo*physicists*. The geophysicists' tools are the gravimeter, the magnetometer, the geophone and the explosive charge. Their vehicles are the ship and the aeroplane.

Sailing or flying over an area repeatedly, the geophysicists attempt to build up a physical picture of the rock contours below by observing movements on the dials of instruments – fluctuations in gravity, changes in the Earth's magnetic field, or variations in the ability of certain rocks to absorb or reflect energy.

Gravity survey: The aim of a gravity survey is to measure variations in the gravitational pull of the Earth. These will depend, at any one point, on the density of the rocks below the measuring device (gravimeter). Different types of rock have different densities – salt, for example, will give a different reading to limestone. The geophysicist draws up a chart with loops and lines on it similar to the contour lines on an Ordnance Survey map – where a line of equal gravity forms a closed loop on the chart, there may well be an oil-bearing rock formation. Today's ship-borne gravimeters are sensitive to changes in gravity of about one part in a million.

Magnetic Survey: Certain rocks are less magnetic than others. Variations, even under the sea, can be plotted to give an indication of the types of rock and their lay-out.

The fastest way of measuring is to use an airborne magnetometer – the sensing instrument is either mounted in the tail of the aircraft or trailed on the end of a cable as a 'bird', as it is called.

Seismic Survey: When an explosive charge is triggered underwater, some of the shock waves travel into the rocks below. They are soaked up or reflected, to a varying degree, according to the depth and type of rock. Knowing the

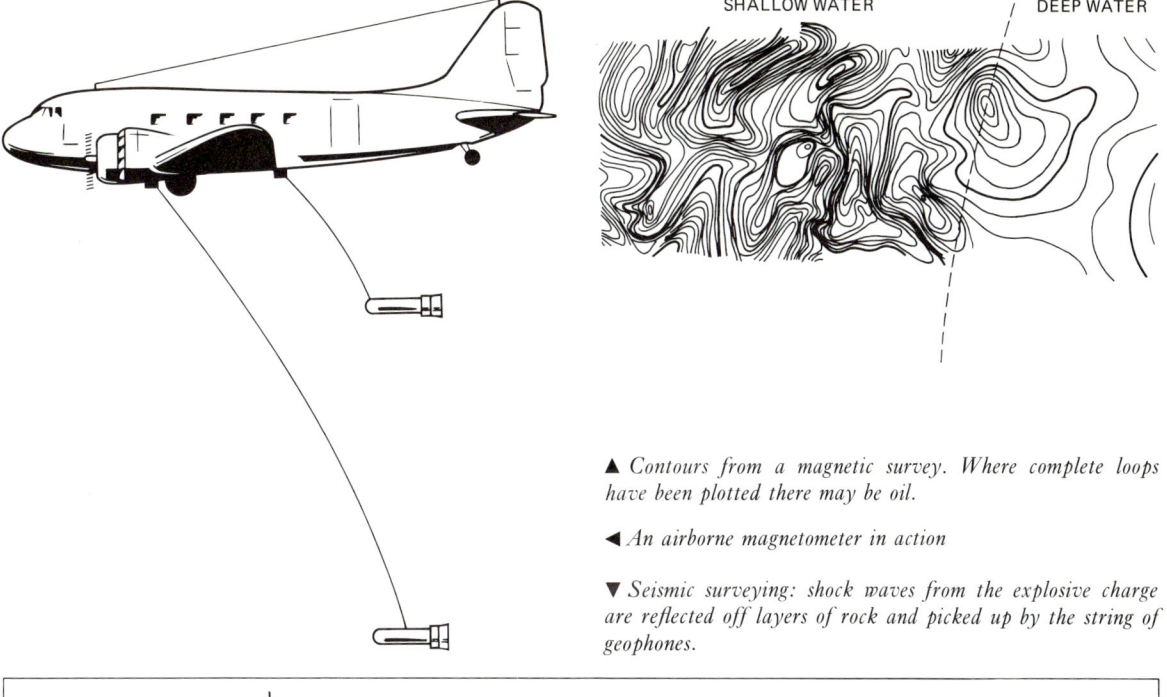

SHALLOW WATER   DEEP WATER

▲ *Contours from a magnetic survey. Where complete loops have been plotted there may be oil.*

◄ *An airborne magnetometer in action*

▼ *Seismic surveying: shock waves from the explosive charge are reflected off layers of rock and picked up by the string of geophones.*

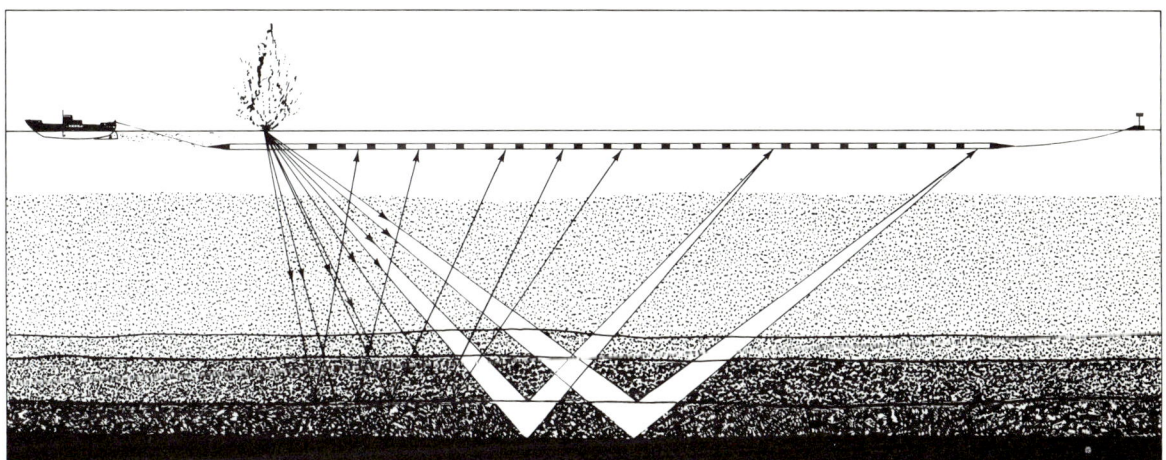

reflective patterns of different rocks – and by measuring the time taken by each signal to reach (and return to the surface from) the various layers, geophysicists can identify them and plot their thicknesses.

In North Sea surveys, ships tow as many as 24 geophones attached to a floating cable. Charges are dropped and fired by remote control. The 'echoes'

from the undersea rocks are picked up by the geophones and recorded on magnetic tape.

Later, computers are used to analyse the data. But final interpretation calls for the skill and judgment of a geophysicist, for much is at stake. It costs about £70 per kilometre to complete a seismic survey – but drilling is an even more expensive business.

Part 4
Drilling

Once, wells were dug by men with spades. Next, engineers developed 'percussive' methods – sharp chisels were alternately raised and dropped. Today all gas and oil wells are drilled by *rotary* action – a 'bit' rotates at the bottom of the hole, crushing the rock, while a fluid is used to flush the chippings to the surface.

The prototype for today's rotary bit was invented by an American, Howard Hughes, in 1909. The procedure for drilling has remained virtually the same ever since.

A derrick is positioned above the spot to be penetrated. Beside it is set a hoisting drum, known as the 'drawworks'. Uniform lengths of pipe (today 30 feet – about 9 metres) are hung from a hook on the derrick. These form what is called the 'drill string'. The string is lowered down through a flat platform, known as the rotary 'table', set in the derrick floor.

An engine rotates the whole table. The string of pipes, with the cutting bit on the end, rotates with it. The weight of pipe, pressing down but always under control, drives the bit deeper and deeper as the bit turns and cuts.

Fluid, known as drill 'mud', is pumped down the inside of the drill pipe, through the bit and then back to the surface via the narrow space between the pipe and the hole which it has drilled. This is known as the 'annulus'. The fluid – originally mud but now a complex mixture of chemicals – not only keeps the cutting bit cool but transports the chippings to the surface.

Bits have improved greatly over the years. Many today are made of tungsten carbide and studded with industrial diamonds. They can be rotated at 250 revs per minute and will penetrate 100 metres of soft rock an hour. High-pressure jets have been introduced between the teeth to help force the chippings to the surface.

Whereas the deepest hole drilled in the 1930s was just over 3000 metres, many today go down more than 8000 metres. Some in the North Sea reach 6000 metres. Steam power

*Drilling bits used in the North Sea.
Some are tipped with diamonds.* BP

10

for the derricks has given way to petrol or diesel engines.

Engineers have learned to line the holes with steel casing, as they go deeper, to prevent caving in. This is cemented into place. And to assist the passage of the drill pipe through the rotary table, they now use a 'kelly' – a square jacket, about 12 metres long, which slides down through the table and rotates with the pipe attached to it.

When the top of the kelly reaches the table, due to the drill having deepened, a fresh section of drill pipe is added. The kelly is then raised by the drawworks once more (hoisted up 10 to 12 metres) and drilling resumes.

As soon as the hole is a few metres deep, the first section of casing is cemented in. This is known as the 'conductor' or 'elephant's ear' and may consist of several lengths of large-bore pipe – perhaps 500 mm in diameter. Its purpose is to provide a flange at the surface.

As the drill bites deeper, more slender casing is used. The drill pipe and bit also become smaller.

A typical exploration well in a new gasfield, for example, might use 500-mm pipe for the first 30 metres, 340-mm pipe for the next 1000 metres, 240-mm pipe for the next 1200 metres and, finally, 220-mm diameter pipe to the bottom of the well. At each stage, the casing is fixed into position by pumping cement down the pipe and up the annulus, waiting for it to set before moving on.

As soon as the gas or oil reservoir has been reached, the well has to be 'completed'. The simplest way of doing this – known as the 'barefoot' method – is to cement the final string of casing into the cap rock above the reservoir, drill through and allow the crude fuel to flow up to a simple well-head (known as the 'Christmas Tree').

The well-head has rams and chokes to prevent a 'blow-out' and valves to adjust the flow.

More complicated methods of completion include installing a fixed section of casing with several tubes inside, each with its own set of plugs and controls.

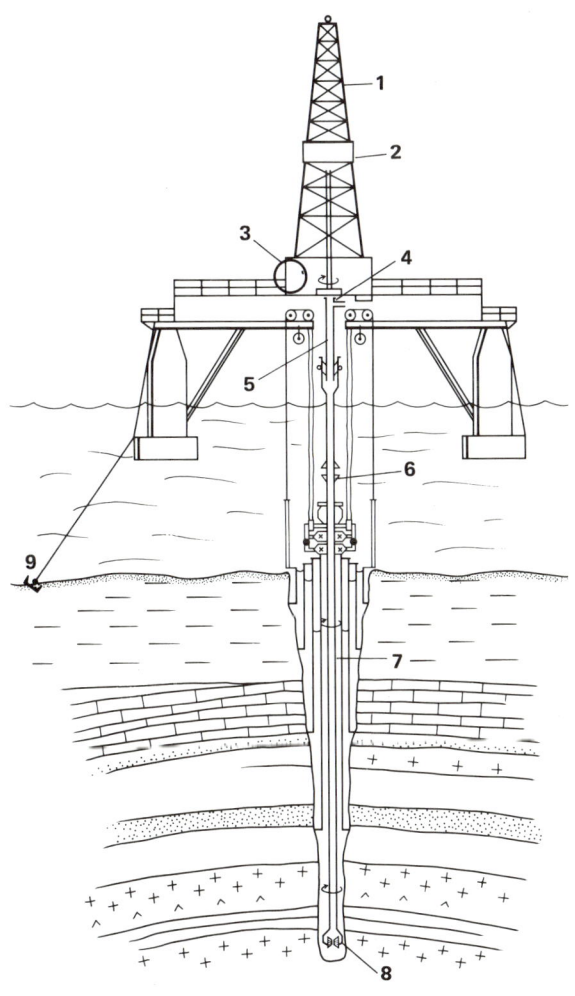

Typical drilling platform:
1. Derrick 2. Hoist 3. Draw-works 4. Rotary table
5. Kelly 6. Drill pipe 7. Casing 8. Drill bit 9. Anchor

Part 5
Readying the Rig

'Steer 270'

The Mariner's command comes crackling over the radio from the Control Room of the drilling rig. Three tugs thrum with 10 000 horse power in response. Heaving powerfully on the lines, they slowly manoeuvre the 7500-tonne steel structure towards a triangle of brightly coloured marker buoys.

'Let go aft anchors.'

Two 15-tonne hooks slip into the sea with a scream of chains.

'Port and starboard tugs take up positions.'

Obediently, two of the tugs fan out sideways until they are holding the rig steady along a north–south line. The third tug keeps its position ahead, making a T-formation into the wind.

'Let go for'ard anchor ... work boats come alongside.'

A pair of flat-decked supply vessels manoeuvre towards the rig. They toss and heave perilously close to the struts and spars, for the sea is angry and the waves hiss with spray. Cranes begin to lower heavy wires, anchors and buoys on to their decks.

There are nine anchors in all and they have to be positioned carefully, each about 1000 metres away and set down in a circle. It is these anchors which must keep the rig steady in all weathers.

Lowering the anchors and positioning the buoys usually takes 12 to 15 hours. But some

North Sea Mariners have waited as long as six weeks to complete the tricky operation because the weather was bad. Care is taken to prevent men being washed overboard by the high seas during anchor laying. If the seas are too high anchor laying does not take place.

'Take the weight' orders the Mariner. Each supply boat reports 'All fast' as the cables tauten.

'Brakeman, let the cable run ...'

Slowly the flat-decked tenders move outwards, drawing cable from the rig like long strands of spaghetti.

'Far enough' radios the Mariner after they have moved about half a mile. 'Put the anchors on the bottom.'

'Anchors on the bottom.'

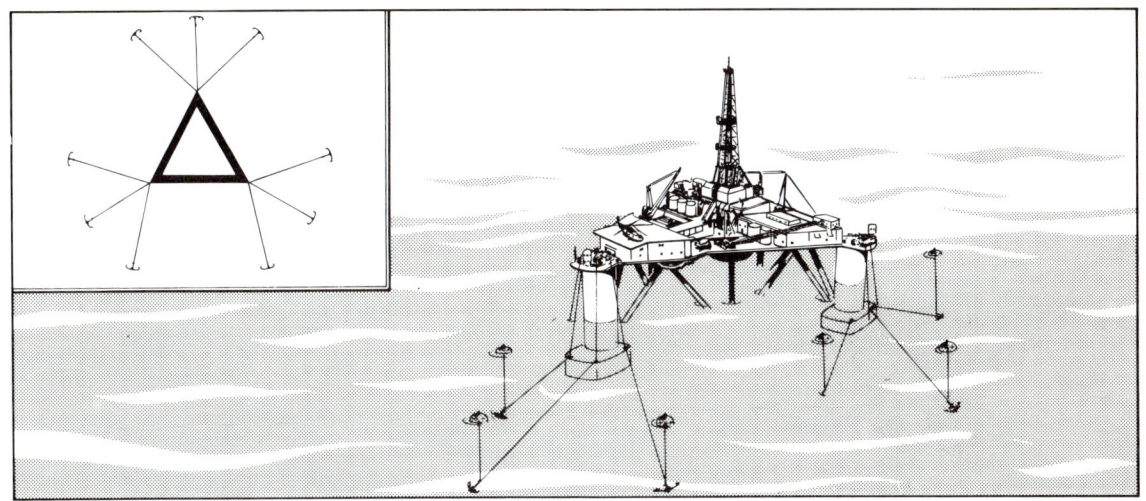

BP's drilling rig Sea Quest is held in position by nine anchors which must hold the rig steady whatever the weather. The anchors are drawn out through pulleys on the legs of the rig and marked by buoys. The small diagram shows the final position of the anchors.

The work boats move out of the way. The rig winchman begins to take up the slack – 50, 80, 100 000 pounds of strain show up on his instruments as each anchor bites into the bottom. Sometimes an anchor drags: back-up anchors, or anchors of different shapes, are lowered.

'OK, Decca. Check position.'

There is an anxious wait while the Decca navigator tender, crammed with radio and electronic gear, moves around the rig checking its position with radio beacons ashore. The captain uses triangulation to get a precise 'fix'.

'Position confirmed.'

The Mariner turns to the Geologist and spells out the rig's exact latitude and longitude. The figures are entered in the log.

'OK,' he turns to the Driller. 'She's all yours.'

On the deck above, steel helmeted men spring into action. This is the moment they have been waiting for. The pumps have been checked. The 'mud' pits are full. Down go the drill pipes, immediately. Running the rig is costing the exploration company £30 000 a day and every minute spent without drilling is £20 wasted.

It is noisy, messy work. As the bit begins to bite into rock and the slurry begins to flow, a brown slime seems to creep everywhere, soiling boots, coats, gloves, faces. Drill bits have to be changed. Extra pipes or casing have to be added to the 'string'. There is vibration from the rotating pipe, clanking from the derricks and echoing thuds from the impacts of waves on the steel supports. The wind can whip a face raw in minutes.

But there is enthusiasm. For *this* well could be the big one . . .

Part 6
Drilling Rigs

Fifty drilling rigs are already operating in the North Sea. By 1980, the total will have reached 100.

There are four basic types of rig or 'drilling platform' – submersibles, semi-submersibles, jack-up platforms and drill ships.

Submersibles, as their name implies, are made to stand on the seabed by flooding parts of their structure. They are rarely used today. They have either horizontal or vertical pontoons below deck – often shaped like beer bottles – which can let in water as ballast. Unfortunately, they can only be used in relatively shallow water – down to about 50 metres. Sometimes, when strong currents are encountered, parts of the structure get scoured and divers have to be sent down to anchor them with sandbags.

Semi-submersibles can be used in deeper water – already to a depth of 340 metres, soon deeper still. They can either stand on the bottom or drill from a floating position. They have large legs with adjustable pontoons at the base.

Some have their own propulsion engines, but most are towed into position. Once on site, anchors are laid out in a circle and pontoons are trimmed, giving the rig stability while it drills.

The largest semi-submersible in the world is the 'Chris Chenery' – 27 000 tonnes, with a deck the size of a soccer pitch and 'legs' (six of them) 69 centimetres in diameter. The

rig was built in Germany. It has its own propulsion system – two 300 kilowatt (400 horse power) motors driving propellers 4 metres wide. Each of its anchors weighs 20 tonnes. It floats on two huge pontoons, 10 metres in diameter, and these are so arranged

British Gas

A jack-up drilling rig. It moves with legs raised. On site, legs are lowered to sea bed and hull is 'jacked' up until well clear of the waves.

that small boats can moor right underneath the rig during bad weather.

Jack-up platforms are basically flat-bottomed barges with legs which can be raised or lowered at will. While travelling to a drilling site, they look rather like rectangular cakes with candles stuck around the edge. On arrival, the 'candles' are lowered and the hull climbs up them until it is well clear of the waves. It is then locked in position.

When the well has been completed, the hull is jacked down the legs and floats again, ready for the next tow.

Jack-up platforms cost less than other types of drilling rig. They can also be readied for drilling faster. For these reasons they are more economical to operate. Most have a limitation, however: the depth of water in which their legs can stand must be no more than 90 metres (recently two new types have appeared in the North Sea which can operate in 110 or 140 metres of water respectively).

Drillships: These have hulls shaped like ships' but with a large hole in the bottom – known as the 'moon pool' – through which the drill passes. They can move swiftly from location to location but tend to bob about a bit.

They are used mostly in deep water, and rely on small thruster engines, rather than anchors, to keep their position. The thrusters are fitted around the hull and controlled by a computer, which calculates the movement of the vessel and switches them on or off – or swivels them – to compensate.

This is known as 'dynamic positioning'.

The largest drillship so far built can drill in 1800 metres of water and is 162 metres long, 10 metres deep and 25 metres across the beam. It can travel at more than 25 kilometres per hour.

BP

A drill-ship – BP's Havdrill. Drillships use small thruster engines to position themselves. The drill moves down through a hole in the hull known as the 'moon pool'.

The world's largest semi-submersible drilling rig – the 'Chris Chenery', built in Germany. Its lower hull measures 115.2 metres × 93.7 metres and each of its six legs is 8.2 metres in diameter. Two 400 hp engines propel it. Eight 18 tonne anchors are needed for mooring.

Part 7
Anatomy of a Drilling Rig

Drilling rigs come in many shapes and sizes but they all have one thing in common – they are expensive. A typical rig for the North Sea might cost £25 million.

They are basically self-contained islands of steel, built to withstand 200-kilometres-per-hour winds and 30-metre waves. They are massive but manoeuvrable, ship-like yet stable.

They are comfortable to live in, but few men live aboard them for very long: most crews spend a week or a fortnight aboard followed by a week or a fortnight ashore.

They have double cabins with bunks, messes for eating in, excellent kitchens, leisure rooms with table tennis and books, shower rooms and cinemas. Most have a helicopter deck to allow men and supplies to be moved on and off swiftly.

Probably the most famous North Sea drilling rig is 'Sea Quest', owned by British Petroleum. It has struck oil on thirteen occasions and drilled more than thirty-six holes.

At the time it was built – 1966 – it was the largest offshore drilling platform in the world, with a triangular deck more than 100 metres long at each side. Afloat it weighs 7500 tonnes.

'Sea Quest' stands on three hollow legs when resting on the seabed. Or she can float and drill with approximately 22 metres of her legs wallowing in the water – nearly 60 000 litres of sea water are sluiced into them to give the rig just the right amount of stability without lowering its hull too close to the waves.

Nine 13 tonne anchors hold her steady while drilling and 18 extra anchors are available in case the seabed is soft.

Her derrick reaches up a hundred feet. Her pumps are driven by three 1.1-megawatt diesel engines, controlling six electric generators, and three cranes swing about her deck moving pipes and equipment. They can lift loads of up to 70 tonnes.

'Sea Quest' can drill a hole 3500 metres deep in up to 200 metres of water. Her tanks are built to hold almost 2 million litres of drilling water, 100 000 litres of fresh water and a million litres of fuel. A distillation plant aboard supplies up to 28 000 litres of domestic water each day – for drinking, washing-up or showering.

Her crew averages sixty in number. They are under the command of a Mariner (captain) or – in the case of those involved in drilling – a Drilling Superintendent or Driller. They come and go by helicopter.

Supplies are also ferried aboard by small boats – work-boats. These may handle anchors, lines, buoys, food or back-up sets of drillpipe. At least two are usually fussing about around a rig at any moment in time.

The drilling team includes at least three senior men – the Drilling Superintendent (who may be responsible for more than one rig), the Toolpusher (responsible for keeping the rig supplied with the necessary materials) and the Driller (in charge of a shift). The rest of the team are mainly 'roughnecks' or 'roustabouts'.

'Roughnecks' work on the rotary drill. They have trade titles like Derrickman. Floorman, or Pumpman – in other words, they assist the Driller. 'Roustabouts' are general handymen – they load and unload equipment, fasten wires, prepare materials and clean up. There are also crane operators, mechanics, electricians, welders, motormen, catering staff and a medic – usually a male nurse.

They usually work 8- or 12-hour shifts, with an hour off for lunch.

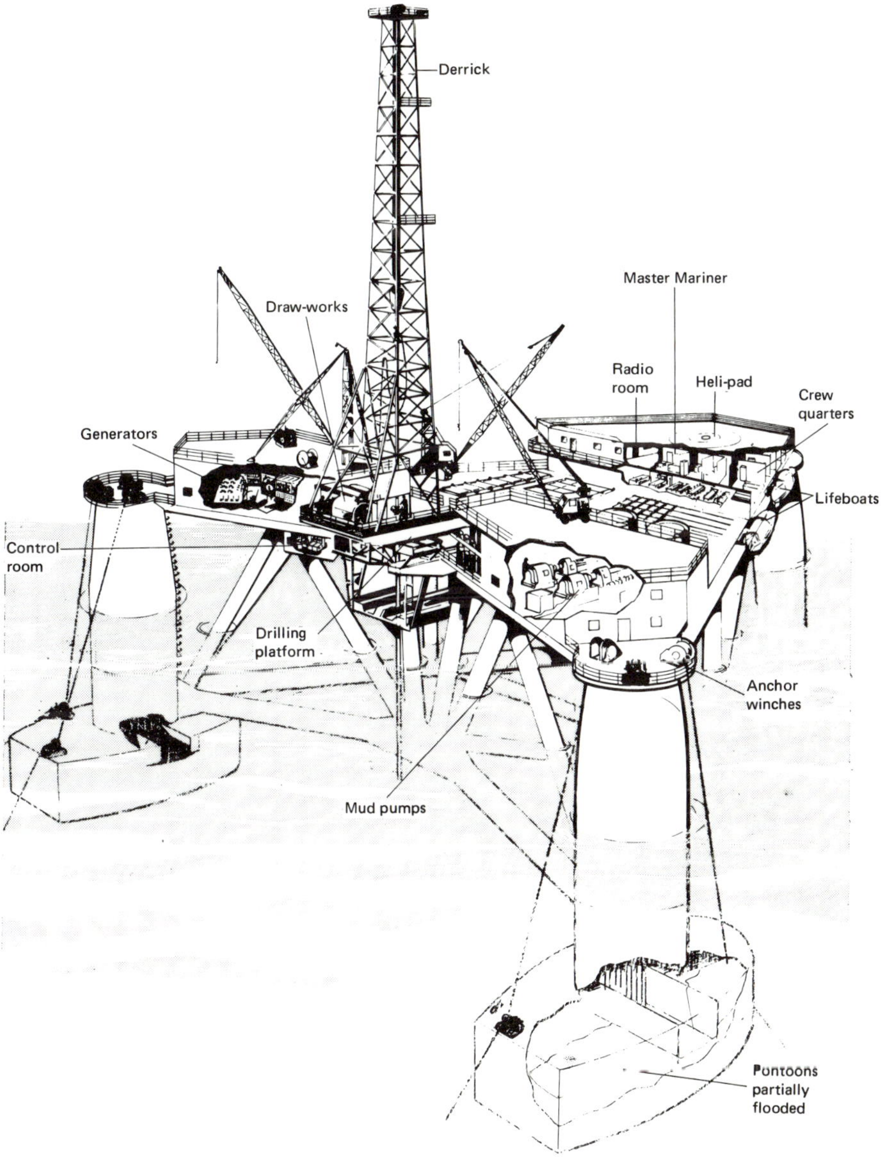

◄ *Inside a drilling rig – BP's £3.6 million 'Sea Quest'. Each of its triangular sides is more than 100 metres long and from feet to top of derrick measures 97 metres.*

Derrick

Master Mariner

Draw-works

Radio room

Heli-pad

Crew quarters

Generators

Lifeboats

Control room

Drilling platform

Anchor winches

Mud pumps

Pontoons partially flooded

▼ *Roughnecks at work on the rotary platform.*

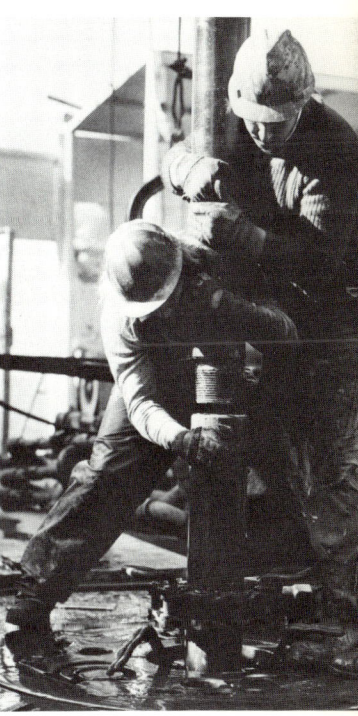

Mobil

Part 8
The Production Platform

Once a drilling rig has found oil or gas, engineers try to estimate the extent of the field.

This is done first by testing the pressure and flow-rate in the first hole and then by drilling 'step-out' wells. Perhaps as many as thirty 'step-outs' will be drilled on all sides around the first hole before the size and geography of the reservoir can be accurately known.

The drilling is usually done *directionally* – that is to say on a slant, at various angles to the vertical. And it may be done not from a rig but from a 'production platform'.

Production platforms are permanent structures which are placed over the heart of a known oil or gasfield. They are the giants of the North Sea. Not all have drills aboard but all have the necessary equipment to control the rate at which the fuel flows from the seabed, to separate oil from gas and to maintain the well-heads and pipeline connections.

There are several styles of production platform in the North Sea but only two basic types – those made of steel and those fabricated in concrete. A third type – a

Mobil

▲ *The concrete legs of Beryl A under tow from Stavanger fjord, Norway.*

Mobil

◄ *The base of Beryl A, one of the concrete production platforms used by Mobil in the North Sea. The three concrete columns rise from 19 concrete cells, some for storing oil, other ballast. A steel deck fits on top.*

▲ *The diagram shows how the latest concrete and steel production platforms compare in size with Big Ben and the Post Office Tower. Notice how small the early production platforms (centre), used in the West Sole Gas Field, seem by comparison.*

Mobil

► *Beryl A in place.*

'hybrid' of concrete and steel – is at the design stage.

The largest steel platforms weigh 57 000 tonnes and look like the Eiffel Tower with its top chopped off. They stand 210 metres tall, in more than 120 metres of water. Their development called for the welding of 7 centimetre thick plates of high-grade steel, careful towing and submerging and the fastening of each structure to the seabed with massive piles.

The piles, up to 1.4 metres in diameter, had to be driven 75 metres below the mudline using the world's biggest hammer. To control the tilting of the towers down on to the seabed, air-filled steel balloons were used and gradually filled with water.

Each tower cost more than £50 million.

Concrete production platforms are cheaper to build. They may weigh a third of a million tonnes and need no pile-driving or anchoring. They simply rest on the bottom. For this reason, they can only be used where the seabed is relatively firm.

The first such 'gravity' platform was built in Stavanger, Norway in 1973 and towed out to the oilfield known as Ekofisk. It is a 237 000-tonne structure, 100 metres high, with huge hollow caverns inside for storing oil. Tankers load up from it. The storage capability means that, if the sea is rough and tankers cannot get alongside, the production platform itself can retain the oil for two or three days until the weather improves.

Part 9
Highland One

The jacket section of Highland One in its flooded construction dock at Nigg Bay on the Cromarty Firth in Scotland.

Some 200 men now live in the part of the North Sea known as the Forties Field. They live on four towers of steel, whose legs go down through over a 100 metres of water – production platforms, known by the places where they were built. *Highland 1* and *Highland 2* were built at Nigg Bay on the Cromarty Firth in the Scottish Highlands. *Graythorp 1* and *Graythorp 2* were built at Graythorp, near Hartlepool on Teesside.

Building these towers – each of which contains more steel than the Forth Bridge and four times more than the Eiffel Tower – was one of the most difficult engineering tasks

ever undertaken in Britain. About 2000 men worked on each.

The great steel sections had to be lifted – calling sometimes for seven cranes, each capable of lifting 80 London buses. They had to be welded, with the welders often balanced 50 metres up in the air. They had to be 'stress relieved', each sub-assembly being baked in an oven to 300 °C. And finally, they had to be cleaned – with every part blasted spotless before being coated with epoxy paint. Then they had to be positioned out in the sea.

Before they could be built, vast dry-docks were constructed so that the towers could be assembled on their sides. At Nigg Bay,

Highland One being towed from the construction dock.

The long journey to the Forties Field.

The jacket section being lowered to the seabed in a controlled crash dive.

1 million cubic metres of sand dune were carved away to provide a 160-acre (60 hectare) dock and site. Each tower was built on a flotation tank so that, when complete, the dock could be flooded and the tank and tower made to rise up in the water. It could then be towed out to sea by tugs.

Flooding a dock took two days. Towing took three days. Final positioning and tilting the tower upright took a further 36 hours. The tilting was done by slowly filling a number of spherical flotation tanks (like giant balloons, attached to the platform's legs) with sea water, whilst keeping others still filled with air. When the angle was right, the whole 34 000 tonne structure was allowed to 'crash dive' to the bottom. Each leg was then pinned down by about 44 50-tonne piles.

The towers are built to last at least 30 years and the whole development of the Forties Field is costing BP about £750 millions.

The crane barge Thor lifting piles on to Highland One to secure it firmly to the seabed.

A helicopter about to land on the completed platform. The deck consists of prefabricated modules, some weighing as much as 2000 tonnes.

21

Part 10
Anatomy of a Production Platform

Fixed, steel production platforms are now producing more than half the oil and gas from the North Sea. They stand on the seabed and have equipment mounted on a large metal box, usually divided inside into two or three decks. The largest can accommodate ninety-six men.

Those built for use by British Petroleum on its Forties oilfield, 175 kilometres northeast of Aberdeen, have three decks. Decks measure 53 by 51 metres. The top deck carries the derrick, four cranes, a helicopter pad, cement and drilling 'mud' storage tanks and a compartment in which the crude oil is separated from its associated gas before being pumped ashore.

The bottom deck houses the 'well compartments' – rooms in which all the oil wells linked to the platform are controlled. Six wells can be controlled from any one compartment, and there are elaborate safety devices to ensure that they all shut down automatically in the event of a fire or gas leak. The middle deck houses the crew.

Each platform has equipment to drill twenty-seven 'step-out' wells at angles up to 55 degrees and depths from 2500 to 3500 metres. This means that each platform is drawing oil from a circle over 5 kilometres in diameter.

The oil is pumped along a 32-inch (812-mm) buried pipeline across 175 kilometres of seabed to Cruden Bay in Aberdeenshire.

Four of these monster platforms are needed to exploit the Forties Field.

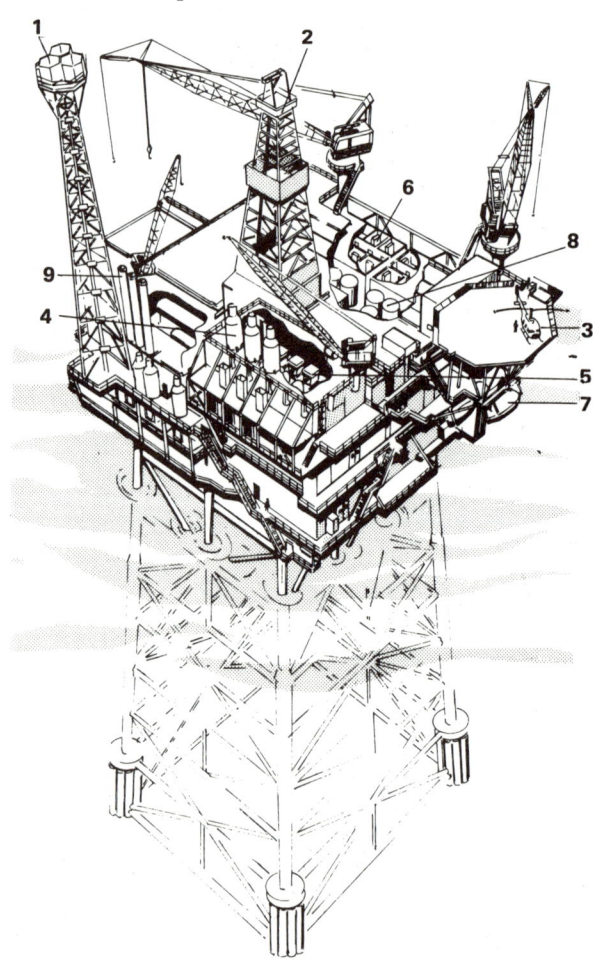

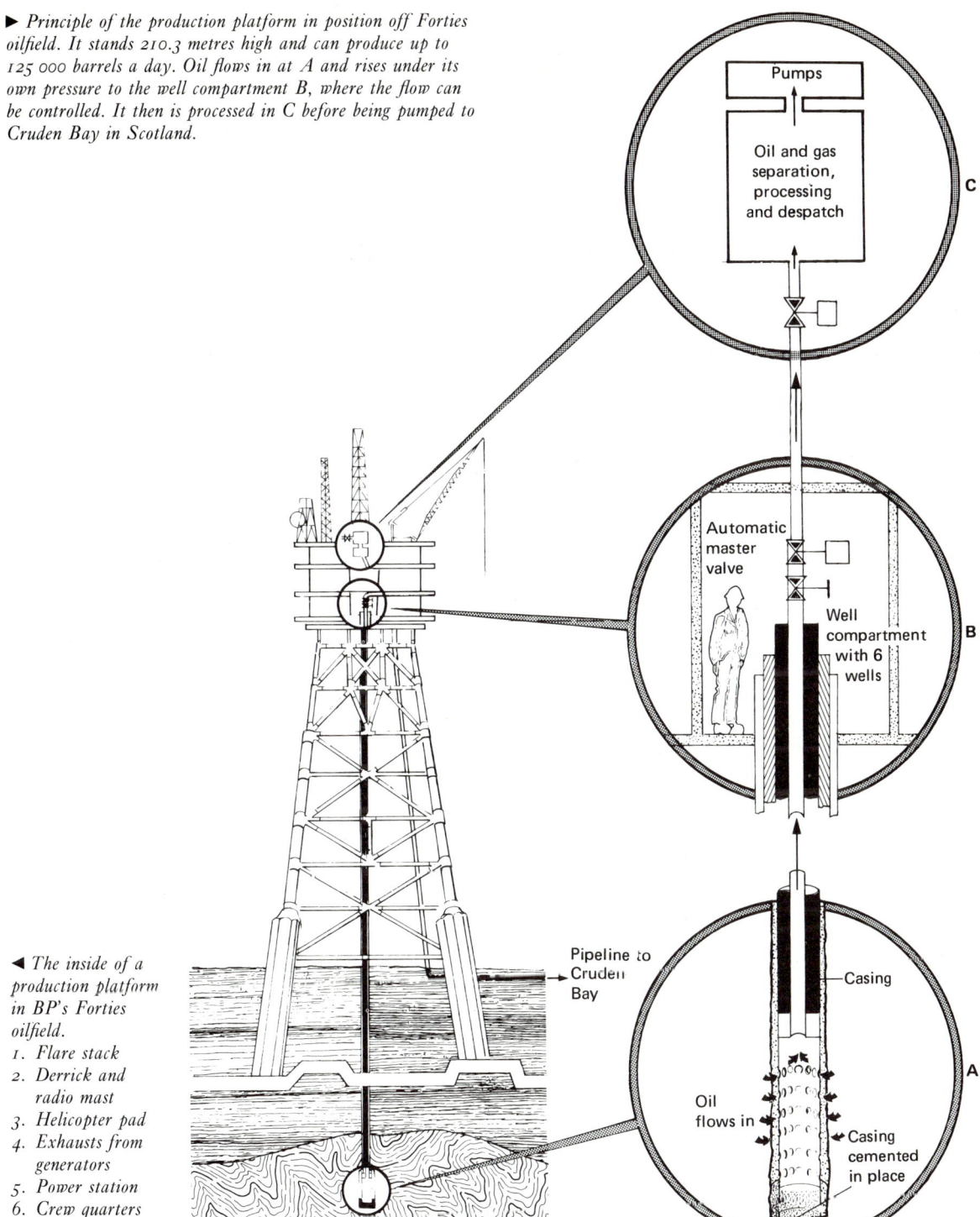

▶ *Principle of the production platform in position off Forties oilfield. It stands 210.3 metres high and can produce up to 125 000 barrels a day. Oil flows in at A and rises under its own pressure to the well compartment B, where the flow can be controlled. It then is processed in C before being pumped to Cruden Bay in Scotland.*

Pumps

Oil and gas separation, processing and despatch

C

Automatic master valve

Well compartment with 6 wells

B

Pipeline to Cruden Bay

Casing

Oil flows in

Casing cemented in place

A

◀ *The inside of a production platform in BP's Forties oilfield.*
1. *Flare stack*
2. *Derrick and radio mast*
3. *Helicopter pad*
4. *Exhausts from generators*
5. *Power station*
6. *Crew quarters*
7. *Lifeboats*
8. *Cement tanks*
9. *Control panel*

Part 11
Man Versus Nature

Life on the North Sea is rough and tough. The forces at work are powerful, especially in winter.

Winds of up to 200 kilometres per hour. Driving rain, snow or sleet. Waves up to 15 metres high crashing against the pontoons or legs of the structures, as often as five times a minute.

Since the exploration for gas and oil started, more than 60 men have lost their lives – swept off decks, crunched by hawsers, felled by pieces of equipment or suffocated on the seabed.

To compensate for this, the men aboard the rigs and production platforms are well paid, well fed and well clothed. Clothing includes khaki two-piece quilted suits (nicknamed by the oilmen 'Chinese People's Army gear'), dungarees, thick sweaters and safety boots. The most prized article of clothing for an oilman is a pair of 'redwings' – thick, leather boots made in Texas and costing about £30.

On the rigs and production platforms, men usually work for seven days, followed by seven days leave. However the crews that man the supply boats work a six weeks on, six weeks off rota.

North Sea crews are paid by the week or the fortnight and *not* paid for time ashore. But it is quite normal for a man to be able to save several thousand pounds in a year, although many North Sea workers go on spending sprees when they *are* ashore.

Most work 12-hour shifts, whatever the weather conditions, and 'stagger' their lunch breaks. A typical menu for a main meal might be: soup or fruit juice; scampi, roast turkey, grilled steak, mixed grill or curried beef and rice; new potatoes, chips, baked beans, broccoli, cabbage, carrots; cherry pie or apple pie and custard; fruit salad, fresh fruit and cheese.

No alcohol is allowed aboard and smoking is only permitted in certain areas. Mail and newspapers are delivered by helicopter several times a week, together with fresh supplies of films or recorded TV programmes.

Most rigs offer table tennis, comfortable armchairs and facilities for hobbies such as modelling or painting. At least one has a strong chess team. Cabins are shared by either two or four men.

The crews who work in the North Sea are of mixed nationalities – British, Dutch, Norwegian, French, German, Spanish, American. One rig even boasts having accommodated a Red Indian chief – Chief Running Bear, whose grandparents were Cree and Iroquoi tribe members. In 1976, he became crew boss on the Sedco 135 rig. Afterwards he commented: 'I've worked for 25 years in the oilfields of the world, but nothing compares with the North Sea. It's one hell of a place.'

▲ *A relief crew arriving by British Airways helicopter on the Shell semi-submersible drilling platform, Stadrill.*

▲ *Relaxing in the mess*

▼ *Crew quarters*

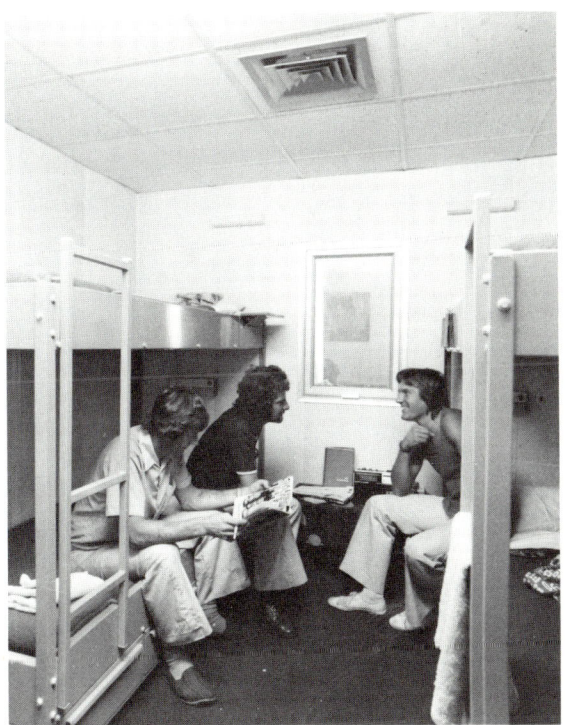

A Shell Photograph

▼ *Big Chief Running Bear – Rig Chief!*

Sunday People

25

Part 12
Call In the Divers

Of all the jobs in the North Sea, there can be none more hazardous than that of the diver. At least twenty-five have lost their lives since the exploitation of the North Sea began.

Divers inspect and install equipment on the seabed, connect pipes and service pipelines. They spend anything from a few hours to a few weeks underwater.

Those who spend weeks below are known as 'saturation' divers. Their bodies become literally saturated with the gases which they breathe.

A diver's greatest enemy is pressure. Pressure increases with depth. Pressure forces the gas which he breathes to dissolve into his bloodstream and, if he returns to the surface too quickly (thus releasing the pressure suddenly) the gas will form deadly bubbles in his joints or brain, maiming him or killing him.

For this reason, all divers who undergo changes of pressure have to 'de-compress' according to a strict time-table as they surface. They pause, at various depths or in a controllable pressure chamber, to allow the gas to seep out slowly. In the early days of North Sea operations, some inexperienced divers did not stick to the time-tables and suffered 'bends' or brain damage in consequence.

Today, all North Sea diving is covered by strict regulations. About 1000 men work under the waves. Some dress in long underwear and heated suits and swim about carrying backpacks of gas cylinders, large face-masks, hand telephones, torches, lifting bags and cutting gear. Others wear light rubber suits. They may go down in diving bells.

Those who go down merely for a few minutes or hours are known as 'bounce' divers. Some of the chambers they use are designed to 'lock' on to a well-head structure: the pressure between the two chambers is equal and the divers can walk through without getting wet and without needing a special gas mixture.

'Saturation' divers are able to stay down for long periods because they breathe a helium and oxygen mixture – no nitrogen as in normal air; this gas mixture does not harm the body while it is dissolved in the blood. They go down for, perhaps, a month at a time. They live in small steel chambers, kept at the same pressure as the water outside, and swim out to work each day on the oil or gas installations.

At the end of a month, they spend several days de-compressing and then go on leave – perhaps for a month or six weeks.

Some North Sea divers are said to be earning £100 a day. But they operate in dark, treacherous conditions, sealed off from the rest of the world and with few home comforts in their habitats, apart from a lavatory and a shower.

At present, the depth limit for 'saturation'

divers is about 300 metres. A variety of submarines and remote-controlled vehicles are being developed for work in deeper waters, and for routine patrolling and inspection of pipelines. They include 'swimming' TV cameras, diving bells with manipulator arms and two-man submarines. The bells and submarines have the advantage that their interiors are sealed and so the men in them can remain at surface pressure and do not need to de-compress.

A new British diving suit, known as 'Jim', does the same thing and is likely to be used in future in the North Sea.

▼ *The diving suit 'Jim' allows man to work 400 metres down at normal atmospheric pressure.*

Lucas Aerospace

▼ *The principle of a 'saturation' dive.*

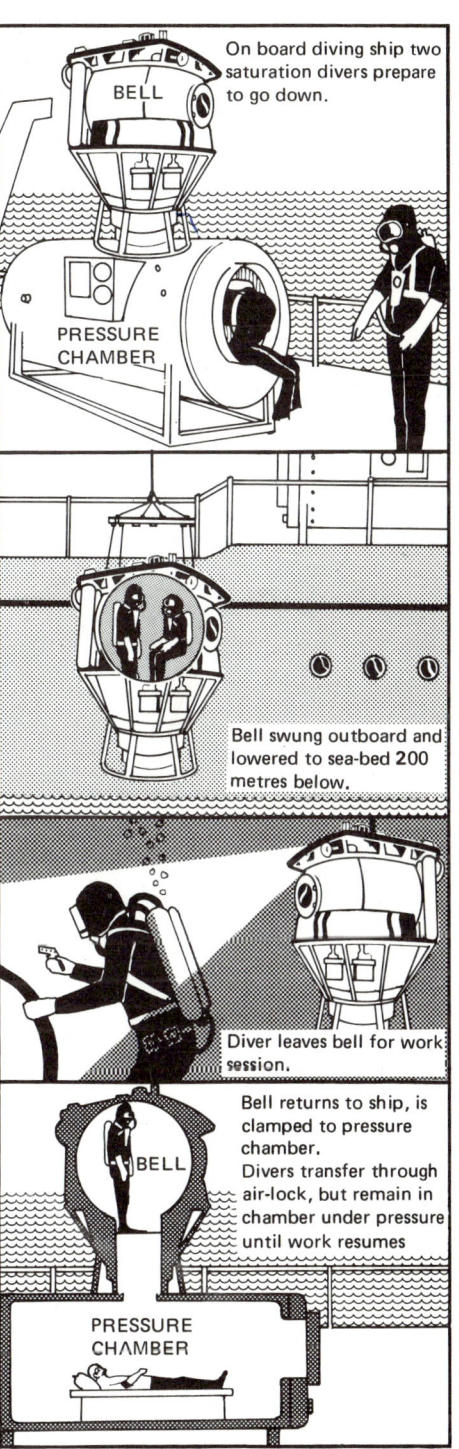

On board diving ship two saturation divers prepare to go down.

Bell swung outboard and lowered to sea-bed 200 metres below.

Diver leaves bell for work session.

Bell returns to ship, is clamped to pressure chamber. Divers transfer through air-lock, but remain in chamber under pressure until work resumes

Part 13
Those Who Service

BP

Omnibus of the North Sea – the helicopter. Helicopters save oil companies millions of pounds a year by transporting men and supplies quickly and in most weathers. They can take off in winds of 65 knots (120 kilometres per hour).

For every man working out in the North Sea, there are a dozen supporting him ashore – or supplying his rig or production platform by boat, helicopter or aeroplane.

Caterers, clothing manufacturers, cleaners, data processors, drill re-furbishers, dry-dock workers, electricians, mechanics, pilots, rope-makers, tugmasters, taxi-drivers, welders – they have tended to gather in groups in towns like Aberdeen and Peterhead, Grimsby and Great Yarmouth, Stavanger in Norway or Rotterdam in Holland.

Aberdeen and Peterhead have mushroomed with warehouses, new harbours and equipment servicing facilities. The Moray Firth and the Shetland Islands have attracted construction firms who have laid down special yards for building the giant platforms – places with names like Nigg Bay, Ardersier and Whiteness Head. Anything up to eighty major rigs are likely to be needed before 1980.

In Aberdeen, Inverness and Easter Ross areas alone, some 12 000 new jobs have been created by the oil industry and a further 5000–6000 are expected, while for Scotland as a whole the total is likely to reach 30 000 extra jobs, although many of these may only be for a limited period.

Some men have made fortunes. Construction workers building some of the giant production platforms have received overtime and incentive bonuses of £3000 for completing their work ahead of schedule.

One man who began with a hot-dog stall near the docks in Grimsby has expanded his business to million-pound proportions by supplying rigs with the right kind of appetizing food.

The aeroplane and the helicopter have become the buses of the North Sea. About 40 000 helicopter flights are made each year, ferrying men and supplies to and from the platforms. Fixed-wing aircraft take personnel between Aberdeen and the Shetlands – or south, on leave.

Helicopters are expensive. An exploration group such as Shell/Esso will spend nearly £5 million a year on transporting men and priority freight by helicopter. But time saved is even more valuable – for example, a trip out to the Brent Field from the Shetlands would take 8 hours by boat (compared with $1\frac{1}{2}$ hours by helicopter) but the journey would probably be so rough that the men would be incapable of work when they got there!

Supply boats, of course, move more slowly but carry more. They are used for ferrying food, machinery, pipes and chemicals to the platforms, for moving anchors or for towing semi-submersible rigs to new positions.

Because the latest North Sea production platforms are so enormous and are expected to drill many 'step-out' wells, extra-large supply boats are coming into service to meet their special needs. The new boats can carry more than 1400 tonnes and have clear deck space of at least 40 by 15 metres.

Most supply boats in the North Sea have two crews and work a rota of six weeks on duty and six weeks off. A typical crew would consist of ten men for servicing platforms or twelve for anchor-handling or towing.

The boats are often at sea in stormy weather and frequently have to run for shelter with their precious cargoes. You need a strong stomach to work a North Sea supply boat.

Finally, there are the safety vessels – one always on 'stand-by' near each drilling platform in case of fire or other emergency. Many of these are 30-metre trawlers with just enough room aboard to accommodate, for a limited period, all the crew of the rig. Recently, specially designed safety vessels of up to 6000 tonnes have been introduced with some of the most powerful fire-fighting equipment in the world aboard.

Workhorse of the North Sea – the workboat. Up to three workboats are usually in attendance around an oil rig moving anchors or ferrying heavy supplies from the mainland.

Mobil

Part 14
Diary of a North Sea Mariner

North Sea Mariners are highly experienced sailors, mature men who have usually seen service aboard tankers before taking command of a drilling rig. Their prime job is to sail and manoeuvre the rig about the North Sea. Once on location, they hand over control to a Drilling Superintendent or Driller. But they remain in overall command, responsible for rig safety, communications and contact with other ships or boats.

Here is a typical day in the life of a North Sea Mariner, once the drill has started to turn:

0600 hrs.	Awake.
0610	Wash. Shave. 'Slop out' – mop down the cabin.
0630	Breakfast. The radio is on, giving the latest news, weather and shipping forecast. 'Gale warnings in the following sea areas, Viking, Forties, Dover, Malin, Sole, Thames.' That means the rig is in for a pasting today.
0700	The night shift comes off duty. The day shift goes on. Go to the Control Room. Look at the state-of-the-rig report – it records stability, what 'mud' tanks are being pumped in or out, what work boats are due to come alongside, what equipment has to be off-loaded and so forth.
0800	Prepare morning report for Head Office. The rig is in a 'tight hole'

situation – no information about what is being found in the hole is to leak out over the air. The company does not want others to know what geological strata they are encountering – whether shale or limestone or what, at the various depths. So it all has to be put into code, before being sent to base by telex.

0830	Call up standby boat. This is always patrolling nearby in case of fire, or if the rig needs to be evacuated.
0900	Call up Head Office on the ship-to-shore radio – telephone. Contact is first made on 2182 megacycles – the radio distress frequency. There is a 3-minute silent period every half hour for distress calls.
1000	Helicopter lands on deck-pad. Mail and newspapers are off-loaded. Two men go ashore, one with suspected measles.
1115	Work boat comes alongside and begins discharging pipe sections.
1130	Order second work boat away to adjust position of No. 6 anchor. Tour and inspect rig.
1140	Divers go down to re-tie knot in cable holding underwater TV camera, showing whether drill has re-entered hole correctly

after fresh 'string' of pipes have been added.

1200   Lunch. Steak, steak and more steak. The drilling crew come in, one at a time, over the next two hours. Everybody sits down together – there is no Captain's Table, or any other sign of rank.

1500   Fire drill. Ring the alarm. Everybody moves smartly to the foam extinguishers, pumps and hoses. Assume the worst and practice 'Boat Stations'.

1600   Afternoon inspection of rig. The weather is roughening.

1700   Safety meeting. Pass on to new members of crew the latest thinking on survival in cold water.

1900   Hand-over to opposite number on night shift. Tell him what boats to expect during the night, the poor weather forecast, and show him all telexes received. Go below, clean up and change.

1930   Supper. Roast chicken or ham and eggs.

2000   Film show. The ten divers (who usually finish work first) have bagged the best seats.

2230   Catch up on paperwork in cabin. Should have done this during the day but was too busy. Weather is really rough now, waves shaking the rig every few seconds.

2300   Go to sleep.

2340   Woken by a terrific 'crack'. Cabin porthole has shattered. So have others. Put on outdoor clothing and make for the Control Room. There learn that an anchor-wire has snapped – its breaking strength was 250 tonnes but it has still snapped with such force that the 'bang' has shattered the portholes.

0300   Manage to get back to sleep, exhausted.

31

Part 15
From the Seabed to the Shore

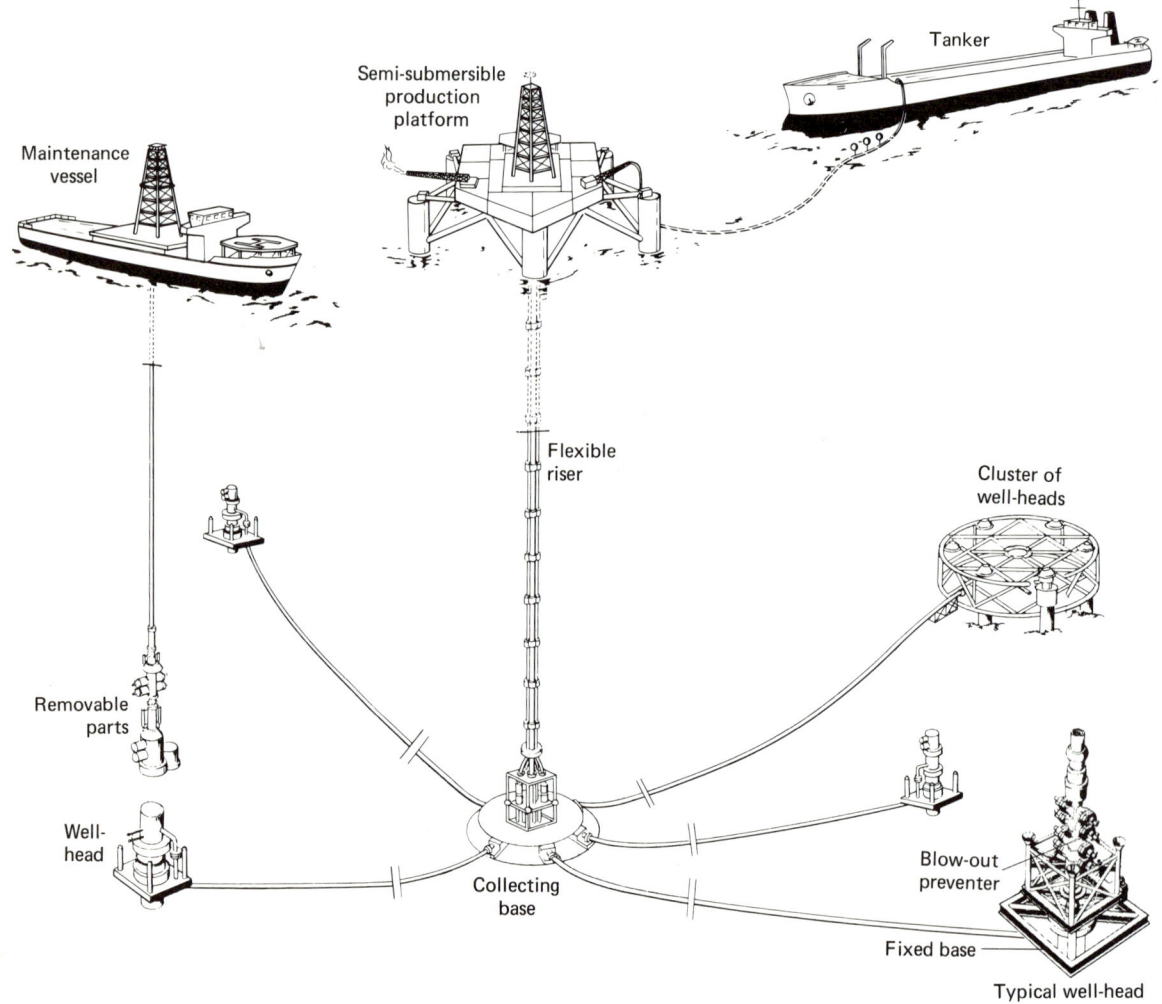

Tanker

Semi-submersible
production
platform

Maintenance
vessel

Flexible
riser

Cluster of
well-heads

Removable
parts

Well-
head

Collecting
base

Blow-out
preventer

Fixed base

Typical well-head

▲ *How well-heads operate in the North Sea.*

▶ *A 'floating petrol pump' – SPAR, positioned in the Brent oilfield. Anchored by six anchors, each weighing 1000 tonnes, it floats and takes in oil from production platforms all around storing up to 300 000 barrels until collected by tanker.*

There are only two ways of getting North Sea oil or gas ashore – by pipeline or by tanker. But there are many different types of well-head for controlling its flow from the seabed. Some are mounted inside a production platform, clear of the waves and receiving supplies from the natural reservoir up a 'riser' pipe. Others are fixed to the seabed itself.

The seabed installations are cheaper but less easy to maintain. They are generally used in deep water. They fall into two basic types – 'wet tree' or 'one atmosphere'.

Wet tree well-heads are coupled directly to a pipeline. They lie exposed to the salty water and are occasionally serviced by divers swimming on to them.

One atmosphere well-heads encapsulate the controls and machinery in a pressurized container: the machinery remains dry and can be inspected and serviced by divers using submersible chambers which 'lock on to' a hatch on the outside of the pressure capsule. The diver walks through, as if entering a room.

Some well-heads are made in two halves so that worn parts can be removed and replaced easily – a cable is lowered from a ship and pulls up the top half so that a whole new section can be bolted into position by a

A Shell Photograph

diver. Other well-heads are joined in circular 'clusters'.

Often, several well-heads are connected to one central 'riser' – a flexible pipe leading to the production platform above. The pipes connecting them are laid out like the spokes of a wheel converging on to a hub. Valves in the 'hub' can be operated by remote control to adjust the flow of oil up the 'riser'.

Pipelines in the North Sea have to be heavily protected. The thickness of the steel used varies from half an inch (12.7 mm) to nearly an inch (25 mm) and many of the pipes are 3 feet (915 mm) in diameter. They are coated with mastic and sheathed in concrete as well.

An alternative to bringing oil ashore by pipe is to connect the wells to a kind of 'floating petrol pump' and then allow tankers to come alongside and load up. Such a 'floating pump' is SPAR – a 140-metre tall cylinder specially built to get oil ashore from the Brent Field before the Shell/Esso pipeline to Scotland was finished. SPAR is anchored in 130 metres of water by six blocks of concrete, each weighing 1000 tonnes. It has six chambers inside, capable of storing a total of 40 500 tonnes of oil. On top is a helicopter pad, living accommodation for twelve, and a turntable. The turntable has pumping equipment and a flexible pipe leading from it and this allows the tanker to swing freely round it to assume the loading position most favourable to wind, waves and current. About 5000 tonnes of oil can be loaded per hour.

As oil is extracted from an underground rock reservoir, so gas or water is frequently pumped back into the empty spaces to keep the pressure up and maintain the flow-rate. But gradually the reserves run out until it is no longer economical to extract the oil. The wells are then sealed.

Part 16
Pipe-Laying

There are already 1400 kilometres of pipeline under the North Sea, bringing the oil and gas ashore, and a further 1200 kilometres are planned. Each kilometre may take anything from a morning to a week to lay, depending on the weather.

Pipe-laying is done by special barges, known as 'lay barges' and 'jetting barges'.

First, the 40-foot (12-metre) lengths of steel pipe are coated – sometimes with glass fibre and pitch, otherwise with coal tar enamel – before being wrapped with plastic tapes. An outer jacket of concrete is fashioned to give protection and also to add weight to the pipe.

The sections of pipe are then loaded on to a flat barge which trails behind the 'lay barge'. The 'lay barge' is a peculiar-looking vessel, with cranes, welding equipment and a sloping ramp at one end (known as the 'stinger') down which the pipe passes on its way to the seabed.

The barge moves forward, one pipe length at a time, by means of anchors. These are usually handled, fore and aft of it, by a tug.

The lengths of pipe are welded together on the deck of the 'lay barge'. Each joint is inspected by X-ray before being coated and wrapped like the rest of the pipe. A zinc bracelet is sometimes placed around every twelfth joint to act as an anode and so protect

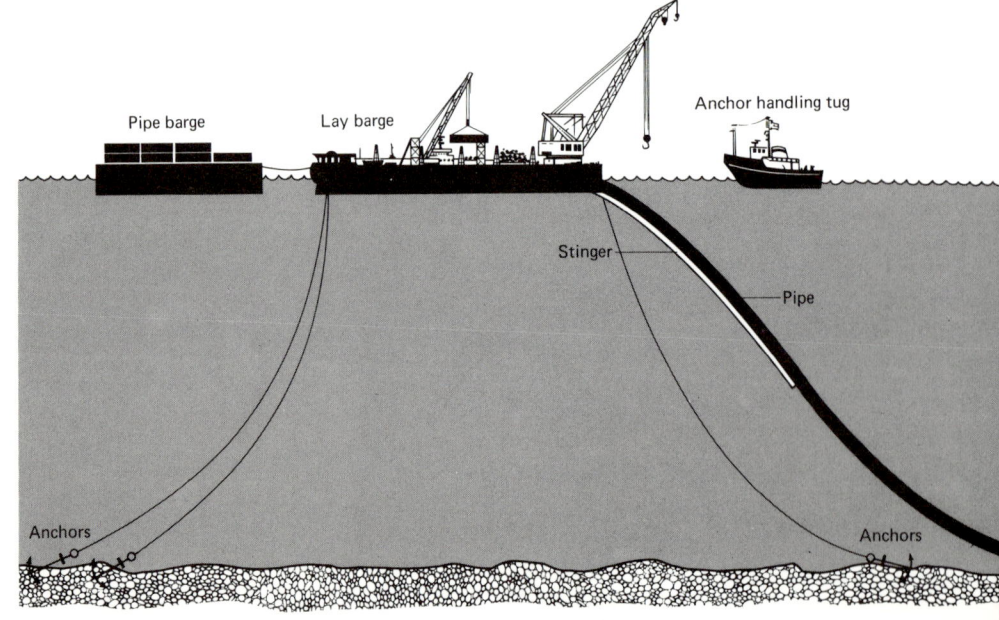

▶ How a pipeline is laid in the North Sea. Note the 'stinger' which takes the weight of the pipe as it is laid to prevent fracture.

Pipe barge Lay barge Anchor handling tug

Stinger

Pipe

Anchors Anchors

the pipe against corrosion. The continuous line of pipe bends a bit as it travels down the 'stinger' to the seabed, but it ends up flat on the bottom.

Another barge then comes along and buries it. This 'jetting' barge blasts out a trench in the mud under the pipe using high-pressure water jets. The loosened material is sucked away. The pipe is then eased into the trench using a special 'sledge'.

Burying the pipe is essential in view of the number of trawlers fishing in the North Sea, whose tackle might otherwise snag the pipe.

Often the pipe-laying operation has to be suspended because of bad weather. One historian, describing the laying of the 70-kilometre pipeline from the West Sole gasfield to the Humber in 1966, commented: 'About 13 000 tonnes of 16-inch diameter pipe were needed . . . in more placid seas, both laying and burying would have been relatively simple, technically. But the operation turned out to be a long, exhausting and

BP

The pipe-laying barge Castoro II during the final stages of laying the pipeline from the Forties Field to Cruden Bay in Aberdeenshire.

extremely expensive battle against foul weather, a sea bed of glacial clay containing large boulders many tons in weight, and even the odd surviving wartime mine.'

That particular pipe-line was laid in record time – a little under nine months.

An alternative to laying pipelines from barges is to weld them together on shore and then pull them into the sea using a powerful winch. Floats are attached to make the pipe buoyant. But this method is only used when the pipeline is short – anything more than 4 or 5 kilometres calls for the special barges.

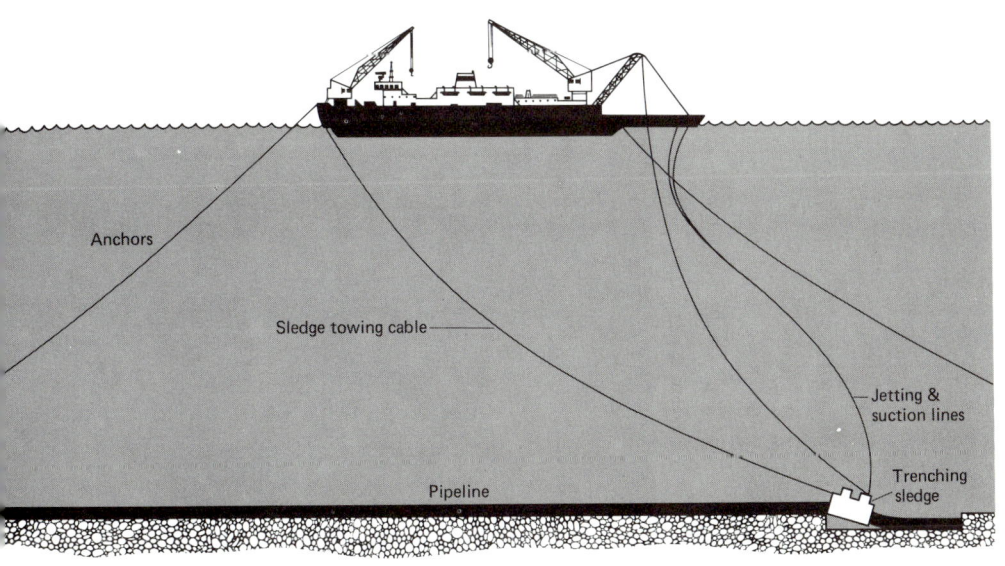

Anchors

Sledge towing cable

Jetting & suction lines

Pipeline

Trenching sledge

◀ *Once the pipe is laid it is usually buried in a trench cut out by high-pressure water jets.*

Part 17
The Oil Comes Ashore

In October 1970, the BP drilling rig 'Sea Quest' found the first major oilfield in the British Sector of the North Sea. It lay in what is known as the Forties area – so called because much of the seabed there is 40 fathoms deep.

The well struck oil 2100 metres down. The field turned out to cover about 90 square kilometres and there would appear to be around 4400 million barrels (587 million tonnes) of oil lying there.

Since then eighteen other oil fields have been discovered in British waters and others are likely to be as the exploration goes on.

The oilfields have been given attractive names – Auk, Andrew, Josephine, Beryl, Maureen, Alwyn and Claymore, Argyl, Montrose, Piper and Thistle, Cormorant, Dunlin, Hutton, Brent, Ninian and Frigg.

Two major pipelines have already been laid and are bringing the oil ashore. The first, 32 inches (812 mm) in diameter, runs 175 kilometres from the Forties Field to Cruden Bay in Aberdeenshire. The second is an even longer pipeline and carries Norway's Ekofisk oilfield products ashore to Teeside for shipping back to Norway (or export) by tanker.

The reason for piping the Norwegian oil to Britain is that a huge, deep trench lies between the Ekofisk Field and the Norwegian coast – too deep and costly to traverse by pipeline – whereas the water on the British side is comparatively shallow.

Five more major pipelines are either under construction or planned. One of these – a 36-inch (915-mm) pipe running from the Brent, Thistle, Dunlin, Hutton and Cormorant oilfields across 150 kilometres of seabed to Sullom Voe in the Shetlands – is costing £200 million. By 1980, it should be bringing ashore about a million barrels (134 000 tonnes) of oil a day.

New terminals are having to be built ashore to handle the oil. At Flotta in the Orkneys, the Occidental Oil Company has constructed five storage tanks to receive oil from the Piper Field, each tank having a capacity of half a

BP

A pipeline comes ashore.

36

million barrels. All such new structures, massive though they may be, are being hidden away or camouflaged as much as possible to preserve the look of the country-side.

North Sea oil is what is known as 'light' – it has to be mixed with heavier crude oil in order to make certain products. But it is low in sulphur, which makes it attractive for burning to provide heat and power.

All crude oil is a complex mixture. All North Sea oils are complex mixtures – no two fields yield exactly the same. Each refinery has to be adjusted to meet the individual mixture but the end products are roughly the same – gasoline, aviation fuel, diesel oil, heating oil, lubricants, road asphalt and a host of basic ingredients for the chemical, plastics and fertilizer industries.

All these are beginning to come from the British refineries handling North Sea oil – 24-hour 'fuel factories' at places like Grangemouth in Scotland or South Killingholme on the Humber. But there is one unusual product coming out of BP's refinery at Grangemouth – protein for animals to eat.

In the 1960s, scientists developed a process for 'growing' bacteria on oil – the bugs literally fed off substances in the oil, grew fat and multiplied. After treatment, their decomposed bodies could be made into a cake which could be fed to animals as a supplement to their normal diet.

The first 'North Sea hens' have already laid. And the first 'North Sea pigs' have already ended up as bacon.

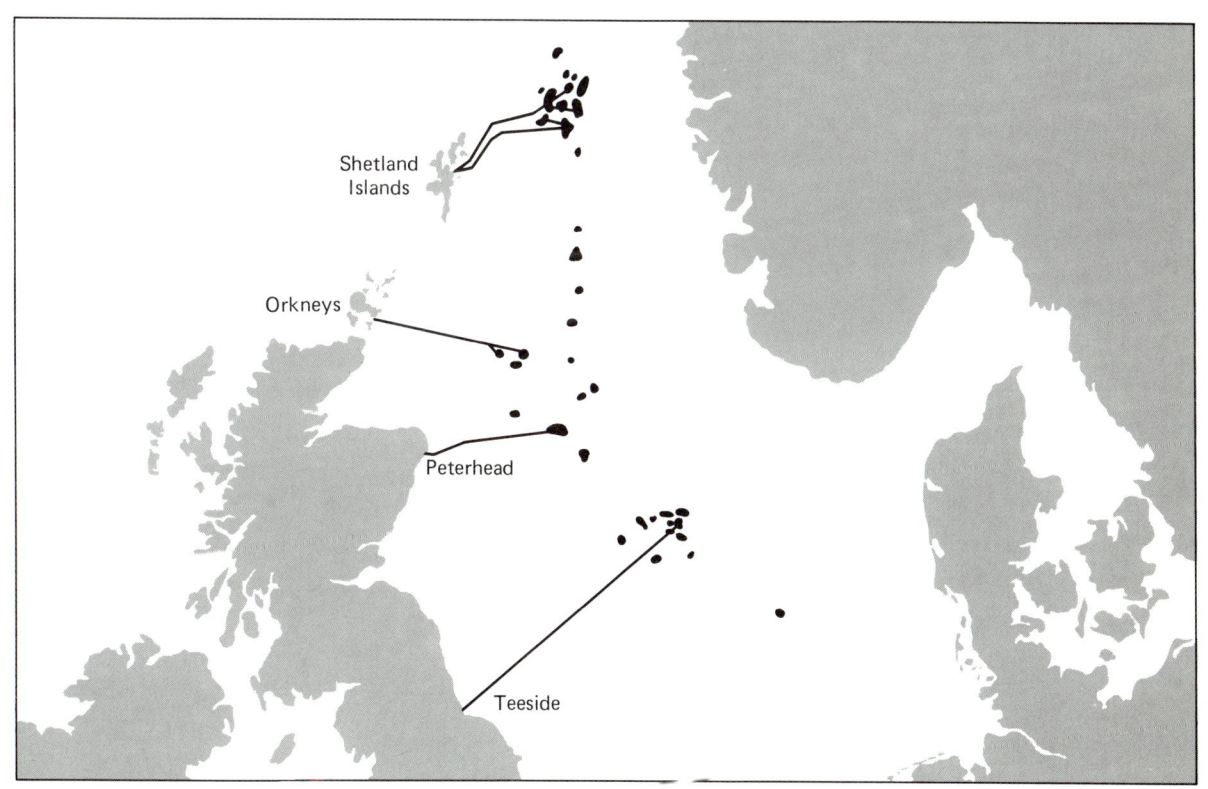

The pipelines so far constructed in Britain's sector of the North Sea.

Part 18
The Gas Comes Ashore

Gas is found in two forms – with or without oil. When found with oil, it is known as 'associated' gas and usually contains a lot of impurities, which have to be extracted before it can be used. 'Natural' gas tends to be cleaner.

Most of the natural gas in the North Sea has been found in the south. It comes from fields with names like West Sole, Viking, Hewett, Indefatigable, Leman and Rough – or, in the case of Holland's gasfields, 'blocks' with numbers like L10 or K13. But in 1971, a drilling rig of the Petronord Group discovered gas in the *north* – in a region known as Frigg, 320 kilometres north-east of Aberdeen. It was a major find. Frigg has turned out to be the largest gasfield ever discovered at sea.

Already it has boosted Britain's energy reserves by 7 per cent, and natural gas reserves by 30 per cent. And before 1980, Frigg is expected to be supplying about 700 million cubic feet of gas a day.

The gas is good and clean, as is all North Sea natural gas. It contains little sulphur. Nevertheless, it still has to be processed.

Some of this processing is done out at sea, on the production platforms. First, any loose water, solids or hydrocarbons are removed from the gas. Then it is dried to prevent the formation of methane hydrate, a snow-like substance which might block the pipeline.

Sometimes methanol is added. And finally the flow is metered and regulated, before being pumped ashore.

On the way along the pipeline, the gas sometimes drives along a 'pig'. This is a heavy, spherical object used to clean the pipe. Liquid condensate sometimes gathers in the pipeline and comes out in blobs known as 'slugs': these are caught in a specially tilted pipe at the shore terminal known as a 'slug catcher'.

The gas is then ready for chilling. Chilling liquefies it – thereby reducing the amount of room needed to store it by about 600 per cent. The condensed liquid is separated into two 'fractions' – condensate and wet methanol. The condensate is pumped to a refinery. The wet methanol is taken away by road tanker for use again at sea.

At the refinery, the gas is filtered, to remove dust and then blended with other gases to maintain a standard quality and thermal efficiency. At this stage – as at sea – it has no smell. So smell has to be added, in order to make sure that the public knows when there is gas about.

Sulphur compounds are added in minute quantities to give it smell.

The gas is then ready for pumping into the National Grid. This is a high-pressure network of pipelines over 200 000 kilometres long supplying natural gas to all of Britain. High-strength steel pipes are used, up to 48

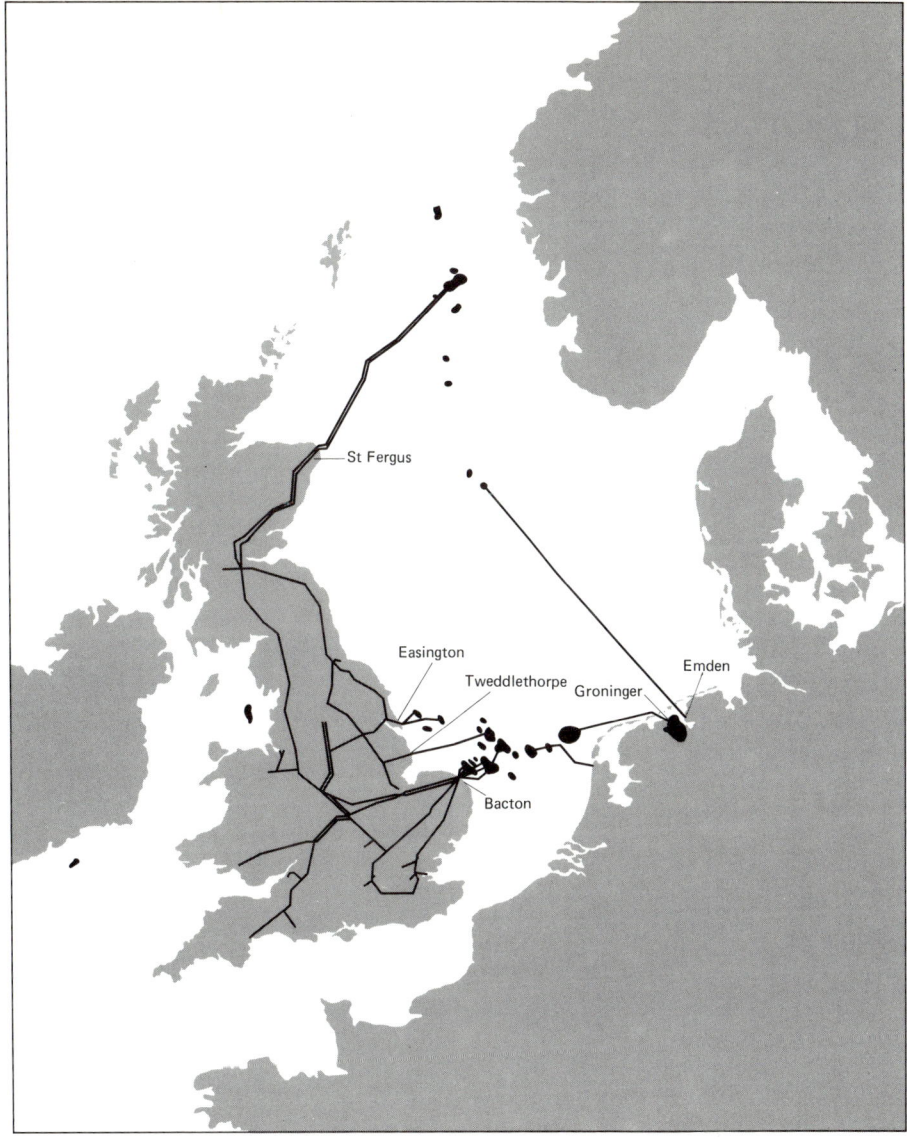

St Fergus

Easington

Tweddlethorpe

Groninger

Emden

Bacton

inches (1200 mm) in diameter. Elaborate steps are taken to safeguard against fracture and the pipes are regularly inspected from *inside* by sending along a mechanical 'pig' with instruments aboard.

A number of compressor stations have been built to boost the gas around Britain. There are also numerous 'reduction stations' to reduce the pressure before it gets to the user.

By 1977, all of Britain's 14 million gas users had been converted to North Sea gas – all appliances had to be altered to work off gas with twice the heating efficiency of previous 'town' gas. Despite the cost of this operation and the cost of laying new pipelines (up to £625 000 a kilometre at sea and about £130 000 a kilometre on land) North Sea gas is now saving Britain about £1000 million a year on its import bills and will go on doing so for a long time to come.

Part 19
Blow-Out

The thing which oilmen fear most is a 'blow-out' – an uncontrollable escape of oil or gas from the reservoir below. Such an accident occurred, with dramatic force, at 9.30 p.m. on Friday, March 21st 1977.

It occurred as engineers and roustabouts working for the Phillips Company were making adjustments to Well No. 14 aboard the production platform Ekofisk Bravo, 350 kilometres south-west of Norway.

Some months previously, the engineers had been making a routine test on the well, using a tubular device for measuring the flow of oil which they lowered down the well. Unfortunately, it slipped off its cable. They fished for it but failed to locate it.

Shortly afterwards, oil began to flow less freely up the well than usual and the engineers suspected that the lost instrument was jamming the pipe. They decided to remove all 3000 metres of pipe and were just about to do so when the 'blow-out' occurred.

(OSL-3)OSLO APRIL 23.(AP)-EKOFISK BLOW-OUT DISASTRE = A fire-fighting vessel spurting water at the "BRAVO" platform after the uncontrolled oil and gas blow-out hit Norway's big Ekofisk oil field in the middle of the North Sea friday night. (AP-WIREPHOTO) oc/stf-str. 333. 1977. SWEDEN OUT   SWEDEN OUT   SWEDEN OUT   SWEDEN OUT.

A fire-boat spraying the Bravo platform with water, to cool the rig and reduce the risk of fire.

A jet of oil and gas shot nearly 50 metres into the air, scattering the men working near the top of the pipe and showering them with filth. The stench of hydrocarbons was nauseating.

The engineer in charge sent a distress call to Norway and immediately tugs and fire-boats closed in on the area, finally positioning themselves upwind of the rig so that their powerful hoses could spray water down on to the column of escaping oil and gas, cooling the whole rig and reducing the risk of fire. The crew were evacuated.

Oil was escaping at a rate of 4000 tonnes a day, forming a huge slick in the North Sea and threatening to pollute hundreds of miles of Norwegian, Danish, German and even British coast-line. Phillips decided to call in the world's greatest expert on 'blow-outs' – a stocky Texan called Paul 'Red' Adair.

Adair was actually working on another 'blow-out' – in Mexico – at the time, but he sent his two lieutenants, 'Boots' Hansen and 'Toots' Hatteberg, by plane to Norway immediately. A support ship, *Choctaw 1*, was quickly moved alongside the Bravo platform so that they could use it as a base.

There are many different ways of dealing with a 'blow-out', ranging from allowing the oil and gas to burn out naturally, to drilling afresh into the reservoir to relieve pressure – or even 'snuffing out' the escape using explosives. Hansen and Hatteberg decided to use a mechanical method.

At the time of the escape, roustabouts had removed Well 14's 'Christmas Tree' – the device, like a fire hydrant, which permanently caps the well and controls the flow of oil up the pipe. They had been trying to replace it with a 'blow-out preventer' – another mechanical device, with metal rams, which can instantly shut off any sudden up-surge of oil if the drilling 'mud' (pumped down to restrain the oil while sections of pipe are removed) proves insufficient.

Unfortunately, the blow-out preventer had somehow been put on upside down – and failed.

For two days, Adair's men worked out their plan and waited for special equipment to arrive. Their mission was hazardous because one spark could have ignited the escaping gas. They used special boots and clothing and worked with brass tools.

On the third day, they tried to close the faulty 'blow-out preventer' – but were foiled by the fact that it was upside down. On the fourth day, they succeeded in bolting a new 'preventer' into position – and Bravo was tamed.

Meanwhile, a fleet of boats equipped with detergent sprays, flexible booms and vacuum hoses were cleaning up the oil slick which had reached 50 kilometres in length and 10 in width.

Happily, the wind kept altering direction – keeping the oil in the middle of the North Sea – and a major disaster was averted.

Thumbs up from 'Red' Adair after his team had finally killed the Ekofisk blow-out.

Part 20
The Economics of the North Sea

Oilmen like to measure in 'barrels'. Gas men measure in 'cubic feet' and 'therms'. Both compare the energy in their product with the energy in coal and so – when estimating the value of what lies under the North Sea – they talk of it having so many 'million tonnes of coal equivalent (mtce).'

Nobody is quite sure how much oil or gas *does* lie under the North Sea. Exploration is far from complete. Estimates keep changing. But in 1977, the British Government's estimate for the UK sector alone was as follows:

OIL		GAS	
Proven	Possible	Proven	Possible
10 125 million barrels	33 525 million barrels	28.7 trillion cubic feet	21.8 trillion cubic feet

In more familiar terms, 10 125 million barrels of oil weigh 1350 million tonnes and, when refined, have the same energy as 2295 million tonnes of coal. A total of 28.7 trillion cubic feet of gas contains 287 000 million therms, or the same energy as 1148 million tonnes of coal.

Put more simply, there is *definitely* enough oil under the North Sea to meet Britain's need for ten years and *probably* enough for twenty: and there is *definitely* enough gas there to last into the 21st Century.

Other nations are not so fortunate. No drilling is being done, at present, in Belgian or Swedish sectors of the North Sea. To date, no worthwhile quantities of oil have been found in the French or German sectors either.

Norway has found large amounts of oil and gas. An estimated 1000–2000 million tonnes of oil and 1–2 trillion cubic feet of gas lie in the Norwegian sector and, even as early as 1975, Norway was producing more oil and gas than she needed and had begun to export the fuels.

Holland too, has oil and gas under her bit of the North Sea. This is in addition to the vast gasfield near Gröningen, whose discovery triggered the whole international exploration programme under the waves. To date, there have been three oil 'strikes' and twenty-seven gas discoveries in the Dutch sector.

Denmark has struck oil and gas also – her Dan Field is currently producing $1\frac{1}{4}$ million tonnes of oil a year, the fuel being extracted by six production platforms and concentrated in an underwater pipeline linked to a 'floating petrol pump' – a mooring buoy – to which tankers can attach themselves.

Altogether, the latest estimates suggest that a total of nearly 50 billion barrels of oil and more than 100 trillion cubic feet of gas can be extracted from the North Sea, over the next twenty years – a bonanza indeed for one of the most oil-thirsty regions of the world, and, for Britain (whose most costly import has been oil), a piece of specially good fortune. Royalties for Britain alone, in 1976, amounted to £66.6 millions – paid by the production companies.

But the North Sea will never rank high in the world table of oil-producing areas. Its yield is never likely to amount to more than 2 per cent of the world total.

Furthermore, the cost of extracting it from the North Sea – £3 per barrel on average – is eight times higher than the cost of extracting it in the Middle East, where much of the oil is on land and relatively near the surface.

Its discovery does, however, give Britain, Holland and Norway a breathing space during which alternative forms of energy – nuclear power, fusion power, water power and wind power – can be developed and made more economical.

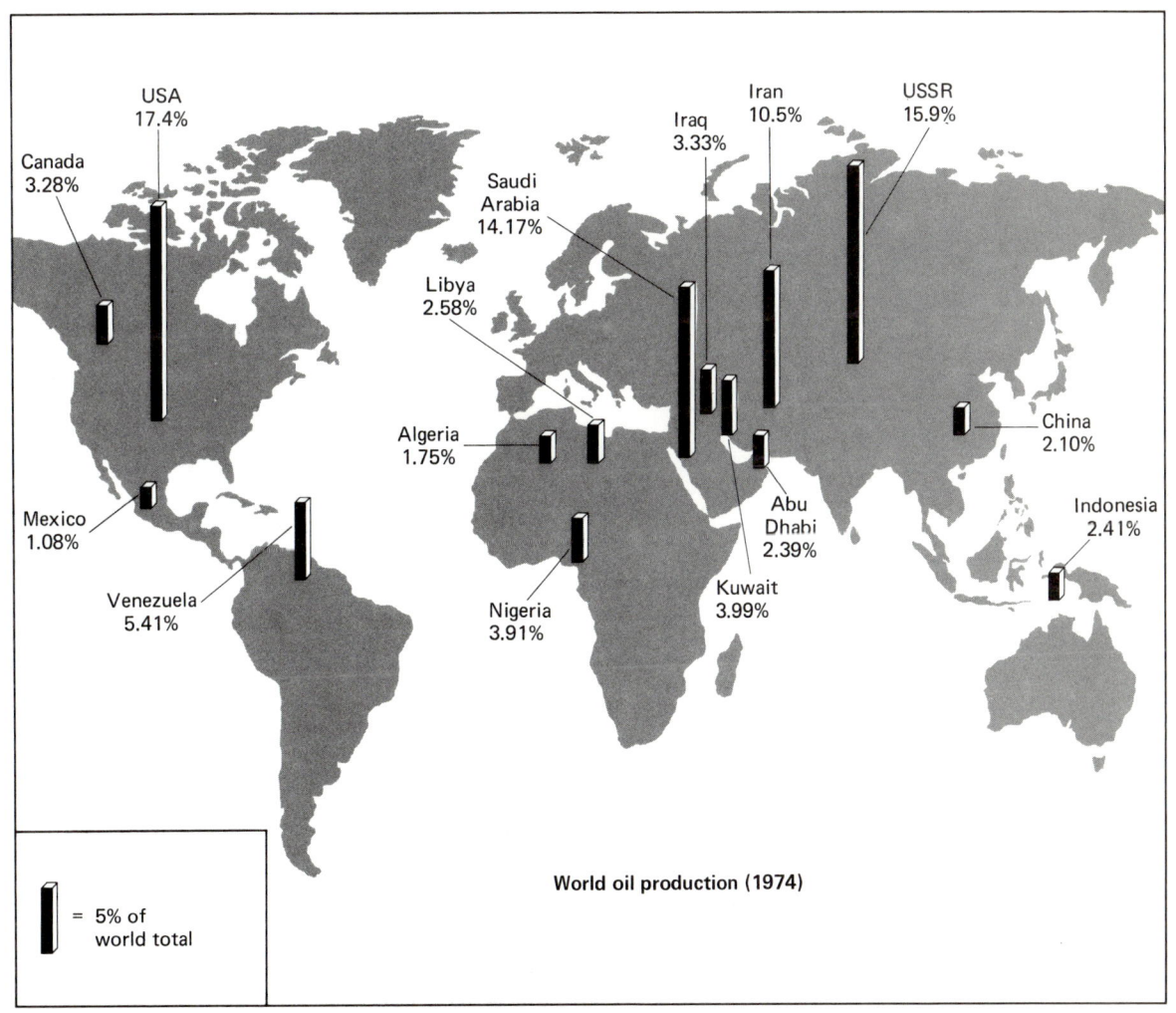

Canada
3.28%

USA
17.4%

Saudi
Arabia
14.17%

Libya
2.58%

Iraq
3.33%

Iran
10.5%

USSR
15.9%

Algeria
1.75%

China
2.10%

Mexico
1.08%

Abu
Dhabi
2.39%

Indonesia
2.41%

Venezuela
5.41%

Nigeria
3.91%

Kuwait
3.99%

World oil production (1974)

█ = 5% of
world total

World oil production by area in 1974. Total production was 14 000 million barrels – or rather more than the present proven reserves of oil in the British sector.

Part 21
The Future

Exploration of the North Sea is far from complete: exploitation is only beginning. It seems probable that there is a lot more oil and gas to be discovered.

The search goes deeper and deeper. The first gasfield was drilled in 30 metres of water. The first major oilfield was in 120 metres. Today drilling is going on north-east of the Shetland Islands in water over 300 metres deep. Soon it will go deeper still.

The deeper you go, the more it costs and the more difficult it gets. Technology is pushed to the limit. All the time, engineers and scientists are looking at new ways of locating and handling the products and cheapening the operation of bringing them ashore.

One idea is to build 'sand towers' – production platforms made of sand. Sand is the most readily available construction material in the North Sea and design studies suggest that if 2 million cubic metres of sand were pumped into a vertical membrane – like a giant sausage standing upright – it might be possible to have a stable drilling and production platform at far less cost than today's steel or concrete platforms, able to operate in water 200 metres deep.

Another futuristic idea comes from Norway – *Conpower*. Conpower is a power station at sea. It is specifically designed for making use of 'associated' gas – the gas which is sometimes discovered with oil. The idea is to build a huge concrete platform and instal a gas-fired power station on top, along with all the necessary drilling and production equipment for extracting 200 000 barrels of oil a day. The power station, according to calculations, would be able to generate 750 million watts and feed the electricity ashore down a submarine cable.

There are new diving suits coming along capable of taking a diver to 600 metres, but still keeping him at normal atmospheric pressure – so that he does not need to decompress at the end of his dive. There are robots under development for routinely inspecting pipelines or well-heads and 'underwater slaves' like the Comex-Seal TOM (Tethered Observation Manipulator vehicle), with claws and TV cameras controlled through an umbilical line.

There are plans for drilling towers which are jointed near the seabed, allowing them to sway and swivel in the wind and waves – permitting drilling in 500 metres of water: and for submerged production systems which can produce oil and gas by remote control.

New ships are being designed and built especially for the North Sea. In fact, a whole new technology is being created.

From Britain's point of view, interest is no longer confined to the North Sea. Drilling has already started in the Celtic Sea, off Wales. Exploration licenses (a licence gives

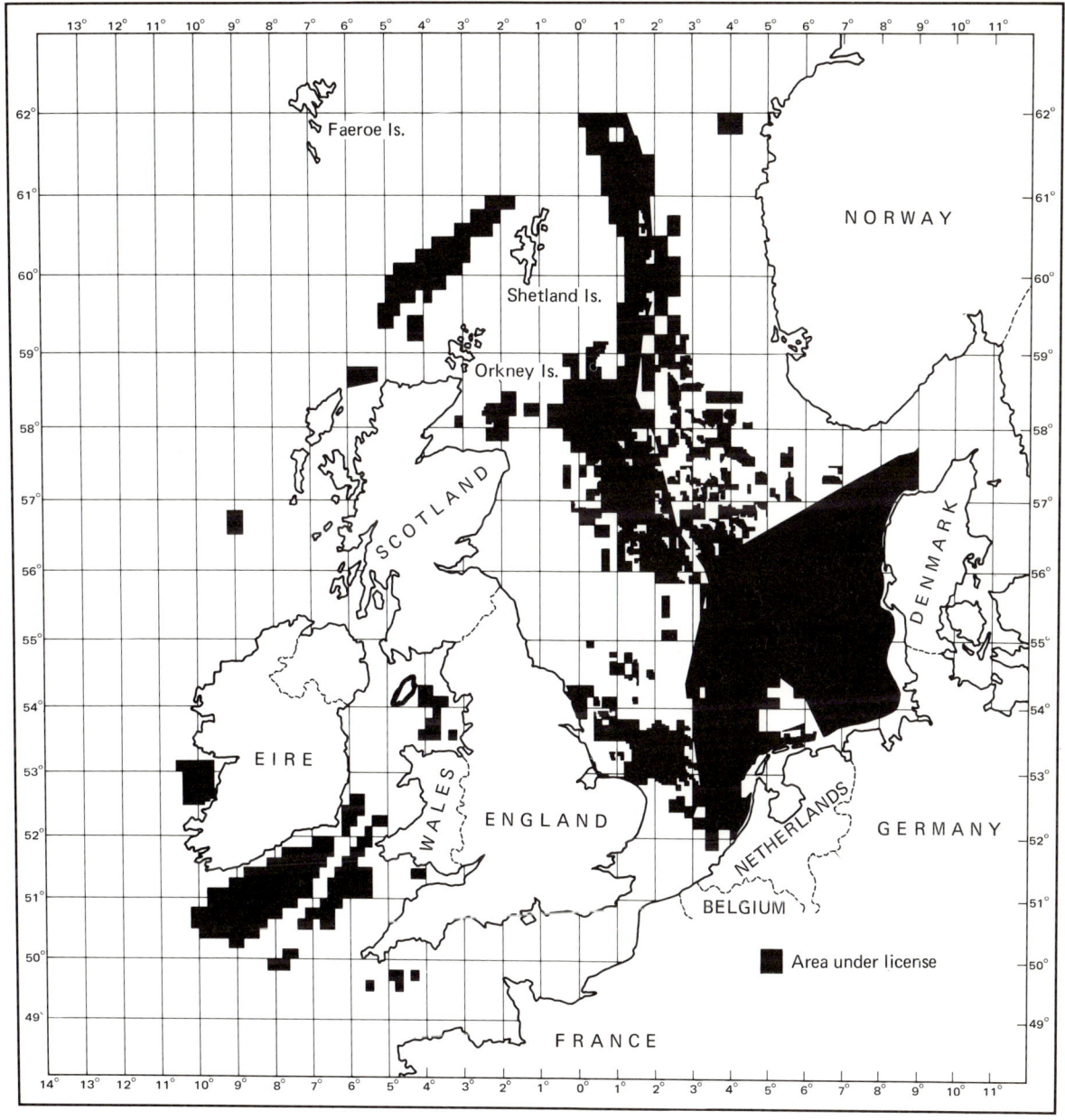

Map of areas around Britain, Europe and Scandinavia so far licensed for drilling. Licensing is done in 'blocks', here shaded black.

the right to survey for oil or gas in a 'block' of sea up to 240 square kilometres) have been taken out for waters off the west of Ireland, the Irish Sea, the English Channel and west of the Orkneys.

As the North Sea reservoirs finally run down in about twenty years time, a fresh collection may well begin spouting oil or gas. The bonanza may be greater than anybody has dared to hope.

45

Glossary of Terms

Anticline: An earth-fold in which the layers of rock have been up-lifted in the shape of an arch.

Associated Gas: Gas found in the presence of oil.

Barrel: A unit of measure used by the petroleum industry equal to 35 Imperial gallons or 159·1 litres: $7\frac{1}{2}$ barrels weigh one tonne.

Bit: The cutting part of a drill.

Blow-out: An unwanted escape of gas or oil from a well.

Butane: A mixture of gaseous paraffins, often known as 'bottled gas' by campers or boat-owners.

Cap Rock: A dense layer, such as clay, which overlies a reservoir rock containing oil or gas and stops it seeping to the surface.

Casing: The steel lining of a well, cemented into position to prevent the well-sides caving in.

Cementation: Fixing casing into position in a well.

Christmas Tree: Nickname for the assembly of valves and connections on a well-head.

Core Samples: Cylindrical slices of rock taken at intervals during drilling and used by a geologist to plot the strata and tell if oil or gas are near.

Crown Block: A metal block mounted on top of a derrick to carry steel lines.

Crude: Oil produced from an underground reservoir after it has been freed from any accompanying gas.

Derrick: A pylon-like frame, usually more than 30 metres high, used for raising and lowering the drill pipe.

Draw-works: A complex set of cable drums used for hauling up or lowering the drill-string, and driving the drilling table.

Drill-String: A line of drill-pipes joined together.

Drilling Table: A flat platform on a rig through which the drill rotates and on which the handlers work.

Dynamic Position: Keeping a rig or ship in position using engines.

Edge-Water: Water lying under a reservoir of gas or oil.

Elephant's Ear: The top section of casing in a well, through which the drill is lowered each time.

Fault: A fracture along which the rocks on one side have been displaced relatively to those on the other side.

Fishing: A term used for the recovery of any object lost down a drill-hole.

Flaring: A method sometimes used to test the quality of gas in a well.

Geophone: A shock-wave recorder, used in prospecting for oil or gas.

Go-Devil: A device for cleaning out the bore of a pipe.

Gravimeter: An instrument for measuring the pull of gravity, used in prospecting.

Holiday: A patch on a pipe accidentally left uncoated.

Hydrocarbons: Chemical compounds of hydrogen and carbon such as paraffins, benzenes and acetylenes.

Inclinometer: An instrument used for checking that a hole is being drilled vertically.